THRIVE

THRIVE

Maximizing Well-Being in the Age of AI

RAVI BAPNA AND ANINDYA GHOSE

The MIT Press
Cambridge, Massachusetts
London, England

The MIT Press would like to thank the anonymous peer reviewers who provided comments on drafts of this book. The generous work of academic experts is essential for establishing the authority and quality of our publications. We acknowledge with gratitude the contributions of these otherwise uncredited readers.

This book was set in Adobe Garamond and Berthold Akzidenz Grotesk by Jen Jackowitz. Printed and bound in the United States of America.

Library of Congress Cataloging-in-Publication Data

Names: Bapna, Ravi, author. | Ghose, Anindya, author.
Title: Thrive : maximizing well-being in the age of AI / Ravi Bapna, Anindya Ghose.
Description: Cambridge, Massachusetts : The MIT Press, [2024] | Includes bibliographical references and index.
Identifiers: LCCN 2023052845 (print) | LCCN 2023052846 (ebook) | ISBN 9780262049313 (hardcover) | ISBN 9780262380188 (epub) | ISBN 9780262380195 (pdf)
Subjects: LCSH: Artificial intelligence—Social aspects. | Artificial intelligence—Psychological aspects.
Classification: LCC Q335 .B365 2024 (print) | LCC Q335 (ebook) | DDC 650.10285/63—dc23/eng/20240222
LC record available at https://lccn.loc.gov/2023052845
LC ebook record available at https://lccn.loc.gov/2023052846

10 9 8 7 6 5 4 3 2 1

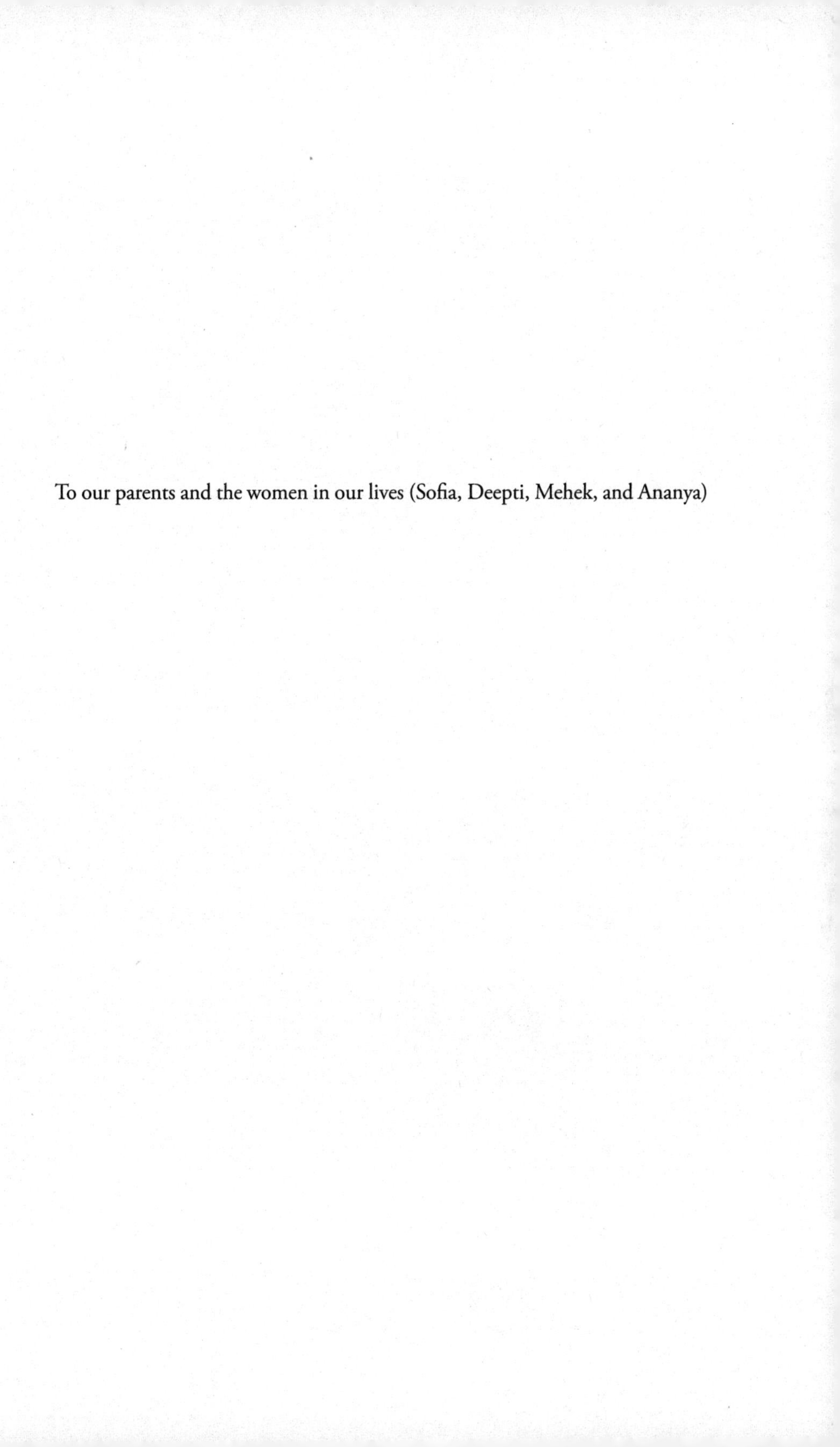

To our parents and the women in our lives (Sofia, Deepti, Mehek, and Ananya)

Contents

Foreword *ix*

1 **EVERYONE IS WELCOME IN THE HOUSE OF AI** *1*

2 **FINDING LOVE** *23*

3 **FOSTERING HUMAN CONNECTIONS** *43*

4 **AI, MHEALTH, AND THE QUANTIFIED SELF** *63*

5 **LEARNING AND EDUCATION IN THE AGE OF AI** *87*

6 **WORK, CAREER, AND FULFILLMENT** *107*

7 **WELCOME TO THE AI-ENHANCED HOME** *127*

8 **CLIMBING THE AI SUMMIT: HOW TO BUILD AI-SAVVY ORGANIZATIONS** *141*

CONCLUSION: MAKING AI WORK FOR YOU *155*

Acknowledgments *161*
Notes *165*
Index *191*

Foreword

In an era where artificial intelligence (AI) emerges as a cornerstone of modern innovation, Ravi Bapna and Anindya Ghose's *Thrive: Maximizing Well-Being in the Age of AI* stands as a seminal contribution to understanding its multifaceted impact on society. AI, as a general-purpose technology, harbors the potential to significantly enhance productivity and address some of the most pressing global challenges. Yet, its complexity, intangibility, and the layers of misunderstanding that often shroud its capabilities and risks distinguish it from earlier general-purpose technologies like electricity and computing. This book cuts through the clutter and demystifies AI in layperson's terms, ensuring a comprehensive understanding of AI's true potential and pitfalls.

Bapna and Ghose, through decades of research and industry engagements, skillfully decode the esoteric realm of AI for the public. By vividly illustrating AI's transformative role in diverse aspects of human life—from education and work to personal relationships—they not only make AI accessible but also empower individuals to engage with it constructively. Their narrative, enriched with real-life examples, serves as a clarion call for harnessing AI's capabilities for societal betterment. It will foster a sense of agency among global citizens to steer AI's trajectory toward enhancing human welfare.

I wholeheartedly recommend *Thrive* to a broad audience—from curious citizens and policymakers to executives and managers. This book is essential reading for anyone interested in fostering informed debates, shaping effective policies, and mitigating the polarizing views often propagated by vested interests. The House of AI framework, presented as a guide through the labyrinth of AI's technicalities, showcases the book's utility as a beacon for all striving to navigate the AI landscape. Its pages promise not just enlightenment but also practical guidance for leveraging AI in crafting a future where technology amplifies human potential and equality.

Nandan Nilekani, Co-founder and Chairman of Infosys
and Founding Chairman UIDAI (Aadhaar)
Bangalore, March 2024

1 EVERYONE IS WELCOME IN THE HOUSE OF AI

March 17, 2018: Mainstream media outlets such as the *New York Times* suggest that "Trump consultants exploited" the "data of millions."[1] This article claims that models (a particular type of artificial intelligence [AI] algorithm that we will unpack later in this chapter) were built to predict psychological profiles of voters and that traits such as extraversion, credulousness, and neuroticism, and interests such as militarism were predicted and then used to craft messaging to psychologically mass persuade voters to swing the 2016 election.

December 19, 2019: As part of a larger effort, a *New York Times* investigation called *The Privacy Project*, investigators provocatively evoke a variety of implications of the advanced (read AI-powered) society we find ourselves in. Once again, this article scares consumers with statements such as "If you could see the full trove, you might never use your phone the same way again."[2]

July 30, 2021: "If You Don't Trust A.I. Yet, You're Not Wrong," say Frank Pasquale and Gianclaudio Malgieri.[3] They allude to Tesla self-driving cars that have crashed, and state that even in areas were AI seems to be an unqualified good, like when using machine learning (also referred to as ML) to detect cancer, we should be worried about whether these algorithms represent all patients' racial backgrounds.

September 20, 2023: AI, specifically Generative AI, regularly passes the Turing Test, exhibiting intelligence that makes it (in many situations) indistinguishable from a human. Generative AI is heralded as the next general-purpose technology that has the promise to do for white-collar knowledge work what IT and automated production, the previous general-purpose technology, did for industrial work by substituting for labor for certain tasks and augmenting human ability for others. Emerging research on Generative AI is showing the promise of increased productivity for elite management consultants both with respect to quantity of tasks completed (12.2 percent more tasks than a control group that did not use Generative AI) and the quality achieved (40 percent higher).[4] Reinforcing this narrative, Microsoft CEO Satya Nadella publicly launches Copilot, an everyday ChatGPT companion that is going to become another app we use on a daily basis along with search, Word, Excel, PowerPoint, and Windows.

As an individual, every day you are being drowned by AI news, headlines and stories from tech companies (which, in the end, want to sell you technology in one form or another), media (which, incidentally, are being disrupted by the tech companies), and, let's not forget, the overnight pundits who can swing from forecasting a dystopian future of AI taking over all our jobs to an overly utopian assessment of a new industrial revolution that will solve all our grand challenges.

So are AI and the digital ecosystem really destined to further impoverish society by exploiting us? Is AI going to make our existing human biases more mechanized and, as some fear, weaponized against us? Will Silicon Valley executives hiring other Silicon Valley executives who design algorithms really make life worse for, say, underrepresented minorities?[5]

Or should we consider AI in the same vein as electricity and computing, more as a powerful general-purpose technology powering the fourth industrial revolution? In this book, we argue that we can also view AI as both a general-purpose technology and a new societal-level operating system that can also have significant positive impact. The negative narrative is one-sided and hence the discussion in the public sphere remains incomplete. This gives them the large-scale ability to affect legislation, shape political debate, and bolster ideologies favorable to their own corporate interests.[6] What is problematic is the one-sided rhetoric against AI and Big Tech that certain entities and some academics have pushed out for the last few years. If you want to binge on more AI dystopia, there is a whole cottage industry out there that will more than satisfy your urge. You can read *Weapons of Math Destruction* or *Algorithms of Oppression*, watch *Blade Runner* or *The Matrix*, or listen to the *CYBER* podcast talking about The Counselor,[7] an AI assistant who (gracefully, actually) escorts and even persuades aging people to their deaths.

To be clear, we are not dismissing the potential downside of adopting AI that has already been discussed at length in many books and articles, including problems related to unfairness, discrimination, and bias. Rather, we are saying that there is a flipside to this public discussion that one should also be participating in, and yet there is a dearth of such a narrative. Why don't we also see AI as the technology that brings shared prosperity and smart solutions to regular people living their dreams and nurturing their aspirations? Why don't we talk about this positive side of AI? In fact, it has been fascinating to watch the public debate on this very topic between the two A. M. Turing Award winners, and the undisputed "Godfathers of AI," Geoff Hinton and Yann LeCun.[8] Hinton has expressed serious concerns bordering on predicting a doom-and-gloom future with AI and even expressed regret for his work,[9] whereas LeCun is much more enthusiastic about a world where AI proliferates to touch many aspects of business and society.[10] We the authors are AI optimists. Our view is that it it's incumbent upon us as citizens and responsible stakeholders to design guardrails around the deployment of AI for the benefit of humanity. We wrote this book to give readers, as individuals, an intuitive understanding and the agency that comes along with it, about what AI is, how it works through the lens of your everyday life, and how you can be empowered citizens in shaping it for your benefit. We hope it will inspire coming generations to pursue careers in AI-related fields and use this powerful technology to go after global grand challenges such as tackling climate change, finding new cures, eradicating poverty and illiteracy, and making the planet more prosperous and equitable.

The first step toward navigating a world of AI is to distinguish between fact and hype. Take the Cambridge Analytica scandal, for example. After the news broke in March 2018, it took less than two weeks for writers at *Nature* to question the "scant science behind Cambridge Analytica's controversial marketing techniques."[11] Within two months, academics at three leading

East Coast universities questioned the validity and effectiveness of the alleged mechanism—psychological mass persuasion—of the Cambridge Analytica scandal.[12] Similarly, Eitan Hersh, an associate professor of political science at Tufts University, told the United States Senate Judiciary Committee on May 16, 2018, that there's little evidence that Cambridge Analytica was, in fact, able to sway the electorate in the 2016 presidential election through its use of Facebook data. Hersh said: "The idea that Cambridge Analytica could use Facebook likes to predict personalities and use those predictions to effectively target ads strikes me as implausible, given what we know about the significant challenges in persuasion in campaigns. No evidence has been produced publicly about the firm's profiling or targeting to suggest that its efforts were effective."[13] But this is not all. Anindya was the testifying expert in the high-profile and nationally watched US government *District of Columbia v. Facebook, Inc.* lawsuit on this topic in which the Judge ruled in favor of Facebook in a landmark summary judgment win. In June 2023 Judge Maurice A. Ross of the DC Superior Court agreed with Anindya and granted Facebook's motion for summary judgment, stating, "Facebook clearly and repeatedly made disclosures to users about its policies such that a reasonable user could not have been misled as a matter of law," and that Facebook users "had numerous tools designed to educate users on their settings and how to protect their privacy. . . . It is difficult to imagine what else Facebook could have conceivably done to be more forthcoming about the privacy settings." The US legal system was smart enough to distinguish between hype and fact.[14]

But where is the widespread media attention for the woman who suffered a seizure in Coburg, Germany? She called 112 (an integrated emergency response number equivalent to 911 in the United States) but could only emit some distressed sounds. Fortunately, the Android phone's emergency location service (ELS) system was still able to send the needed location information to the dispatcher. Emergency responders arrived in time to save her life, but barely.[15] This story was only publicized by Android to highlight the ELS system—no other media outlets ran the story.

Or what about the fact that granular location tracking from smartphones enabled governments around the world to undertake contact tracing and social distancing analyses during the height of the COVID-19 pandemic that potentially saved millions of lives?[16] At least in this case a few scientific and academic publications picked up the story.

What about other positive stories about AI, such as the U.S. Federal Communications Commission estimates that improved location accuracy can increase response time by a full minute, which can save more than 10,120 lives annually in the United States alone?[17] Or that the latest Apple Watch can activate emergency services if its wearer collapses while biking, hiking, climbing, running or playing tennis on a rare 100°F day? (We will unpack *how* it uses accelerometer and gyroscope data together with machine learning to activate these services later in the book.) This feature would have been useful for dehydrated and wiped-out fitness and outdoor enthusiasts in Europe during the summer of 2022 when it went through one of the worst heat waves in modern history.

On the matter of not trusting AI for even ostensibly worthy tasks such as detecting cancer from radiology images, we *should* worry about the root cause of the bias in AI-model outcomes. If the model is more prone to missing out on detecting cancer for, say, Black versus white populations, it is because society does not provide equitable access to healthcare services to these populations in the first place. It's not the technology's fault, it's ours. It should surprise you to know that, in the non-AI world of human radiologists, there is a significantly higher rate of missed detections of breast cancer in images from women belonging to ethnic minorities because doctors are trained using images from white patients.[18] Research shows that when Black newborns are cared for by Black physicians they are twice as likely to survive as compared with white infants because the doctors are more familiar with the specifics of their patients.[19] We live in a complex and inequitable society, and instead of attributing all the blame to biased AI algorithms, it is imperative to look in the mirror and self-reflect. And, as we shall see later in the book and as we teach our graduate students, we can actually create "good bias" when designing algorithms: bias that can *correct* for social frictions and improve the human condition.

So which version of the future of AI should we believe in? Do you feel unprepared to answer this simple question? You are not alone. Unlike electricity and basic computing, the previous general-purpose technologies that powered prior industrial revolutions, AI is more complex, more layered, and fundamentally intangible and invisible to the average Joe. But fear not. The raison d'être of this book is to demystify AI for everyone throughout all aspects of our everyday life. Armed with a greater understanding, we believe you will have agency to make your own judgments about this technology.

Our cards are on the table. We, as academics, are not competing with any company for ad dollars. Unlike influencers, pundits, gurus, or furus (fake gurus), we are researchers who are not racing for followers or fighting for clicks. The harsh reality is that fear and negativity not only sell but sell big. The punditry have their own vested interests in terms of monetizing their negative, one-sided narrative about the perils of AI and technology, so it doesn't behoove them to look at the other side. *Of course*, every technology comes with associated risks—even basic technologies. (Don't take your toaster into the bathtub with you, for example. And drunk driving has never been a valid argument against cars.) Our goal is to arm you with knowledge about how AI works, and to do so in a data-driven, research-based manner. As academics, that's what is in our DNA and what we are trained to do.

At this juncture, it's important to define AI. For the purposes of this book, we are less interested in what is strictly known as *strong AI (also known as artificial general intelligence or AGI)*, where the capabilities of *the* machine are a replica of and indistinguishable from human behaviors. Rather, we deal primarily with *weak AI extending to Generative AI*. This relies on machine learning to perform tasks that are difficult for humans because of a lack of expertise or resources, but that are relatively easy for machines. As an example, think of an AI solution using a deep learning (to be demystified later in this chapter) to read ECGs (echocardiograms) and diagnose heart attacks in real-time in Sub-Saharan Africa where there may be few cardiologists. Or, consider another example, an algorithm sorting through hundreds of dimensions of hundreds of candidates to predict whether a particular individual may be a good date on an online dating platform such as OkCupid, which is a potentially life-altering recommendation since who you choose as your partner is one of the most important decisions you will ever make. We also cover a broader version of AI in the form of LLMs (large language models) such as GPT-4 that can solve a wider range of problems at the expense of having a strong notion of accuracy.

Both authors wear many hats. Between the two of us we are educators, researchers, litigation expert witnesses, tech economists, and data scientists who serve as advisors to venture capital funds and many startups and established companies around the world. We have helped serval hundred businesses onboard advanced analytics and AI and ML projects globally. We have lived and breathed the application of of AI for the last two decades. We have a unique perspective of what AI can do not only for businesses, but also for everyday people going about their lives and for society as a whole.

There are many books about how companies can adopt AI to their advantage. That is not the focus of this book. By contrast, we will demystify AI and make the case that the modern AI-powered ecosystem of digital platforms, apps, and quantified-self devices fundamentally improves the emotional, physical, and material well-being of everyday people across the globe. These will be the stories told in this book.

AI POWERED BY MACHINE LEARNING

Much of the value of AI for most people comes from *machine learning*. Within this branch of AI there are four foundational pillars (descriptive, predictive, causal, and prescriptive) that power everything from online dating to modern health and wellness apps to movie recommendations to networking contacts for job seekers. One common misconception we have noted while advising and consulting for companies internationally over the past twenty years is that machine learning is mostly about prediction and not about causal inference or explanation; it is not about diving deep into the mechanisms of what causes change. This notion is wrong. Causal inference is a big part of AI, as we explain next.

To better understand what machine learning is and its applications, we offer the (fictional) story of Jasmina, a thirty-three-year-old Phoenix-based single mom with a career in fintech as a business development specialist. She's outgoing, connects with people, loves her independence, and, like many of her friends and coworkers, is perennially short on time. Let's look at just part of a week in Jasmina's life.

TYPES OF MACHINE LEARNING AND CAUSAL INFERENCE

Jasmina has been killing it at work, almost saturating the Phoenix region with new adopters of her company's B2B fintech platform. Her boss has recently rewarded her by giving her Texas as an additional region to hunt in for clients. It's the week before Thanksgiving and she needs to compete with the Monday morning rush to catch a flight from Denver to Dallas. Jasmina has an amicable relationship with her ex—her son's father—who is on childcare duty for the week. She lands safely in Dallas, has three back-to-back client meetings, takes a thirty-minute spin on the hotel's Peloton, and she still has an hour to hit the Galleria Mall before dinner with a high-school girlfriend. The endorphins from her bike ride have kicked in and she is reminiscing about a romantic hiking weekend about a month ago with nerdy-but-cute Alex, an also-fictional data engineer at work. The warm and fuzzy feelings trigger an impulse to buy a top-of-the-line Arcteryx jacket for her brother, with whom she has been having an irksome relationship lately. Her mood is rudely sullied by the cashier, who snarkily tells Jasmina that her credit card has been declined.

Why was her card declined? She's nowhere near her credit limit. It turns out that her bank is watching out for Jasmina by deploying a widely used *unsupervised machine learning* algorithm called *anomaly detection*. Her bank has noted transactions of $1,000 and more on her history. It has also noted plenty of clothing purchases, mostly women's clothes. The bank has not, however, seen a $1,000 men's mountaineering jacket purchase—in Dallas, no less—on her account before. It's those three things together—the $1,000 purchase, the fact that it's for men's clothing, and the purchase location of Dallas—that raise a red flag. It was, in fact, her first visit to Dallas. The bank's AI (anomaly detection algorithm) looks at every transaction of Jasmina's and of its tens of thousands of other customers and computes a score that tells it how different or anomalous a given transaction is compared to its "neighbors." It does this in real time and alerts the bank of anything unusual, as was the case for Jasmina, resulting in the declined card.

How can a bank determine which transaction is distant or similar or different from other likely transactions? To do this, Jasmina's bank uses a workhorse concept in machine learning called *multidimensional similarity*. To get to the bottom of this, we really have to jog our brains back to middle-school math. We promise this is the only math we present in this book. Remember the Pythagorean theorem? Whether you do or not, let's look at a different example to see how this works before we return to our declined credit card story.

Jasmina sometimes checks out a few guys on OkCupid or Bumble, her two favorite online dating apps. Alex is cute, but she feels that occasionally she should explore a few more options. A key task for online dating platforms is to recommend dates to users, much in the same way Netflix recommends movies, Goodreads recommends books, and Amazon and Google recommend, well, everything. To simplify matters, let's say our dating app of choice has three date candidates to recommend to Jasmina (see figure 1.1). For these three users, the app has information in three dimensions (again, to keep things simple for now, as in reality, the app will have data on many more dimensions). In this example, we're using income, height, and attractiveness. This

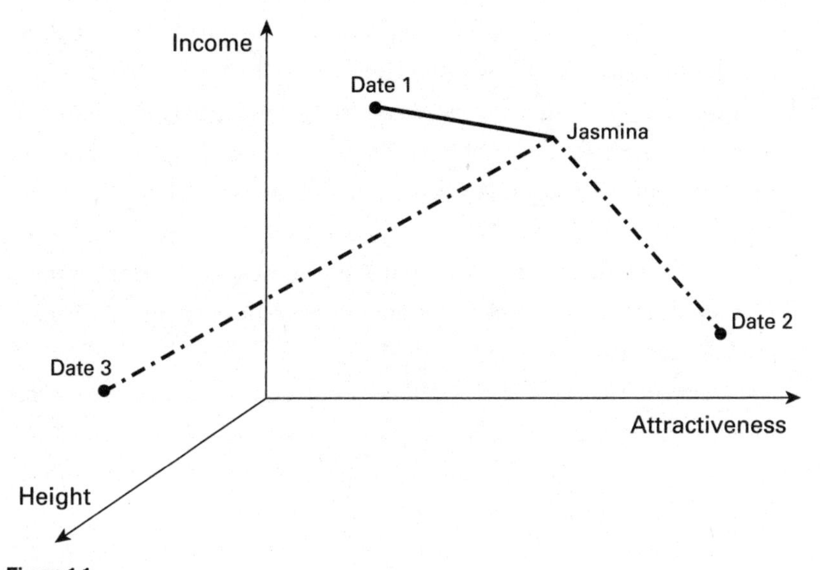

Figure 1.1
Jasmina is "closer" or more "similar" to Date 1 candidate than to Date 2 and 3 candidates.

is similar to our bank, which had data on the category of the item purchased, the dollar amount of the purchase and the location of the purchase.

If the dating app follows the conventional wisdom that birds of a feather flock together, it can mathematically compute the distances between Jasmina and the three date candidates and then recommend the closest, most similar person, which in this case is Date 1. Of course, as we alluded to earlier, a real dating app, just like a real bank, will not have only three dimensions, but perhaps thirty or 300 or even 3,000 dimensions about their users. For instance, every word the date candidate mentions in their online profile, every location where they tag, or every pixel in their photos could be a potential dimension that could be used to compute similarity with other users. But luckily for us, the same middle-school math that allows us to visualize and mathematically compute distance in three dimensions works for 30,000 dimensions.[20]

Armed with the concept of similarity or distance between two points in a hyperspace, we can be very dangerous (read: useful) with our application of machine learning. (Allow us to geek out a bit by introducing the concept of *hyperspace*, which is something more than a two-dimensional or a three-dimensional Cartesian graph type space that we can easily visualize. In this case, perhaps a thirty-dimensional version of the three-dimensional figure 1.1). For instance, let's visualize Jasmina's intended purchase of the jacket for Alex in a similar three-dimensional visual (figure 1.2).

Clearly, Jasmina's intended purchase, the star-shaped point, looks far away from its three closest neighbors. By contrast, another recent transaction, as shown by the diamond sh-ped point, does not seem so different from other transactions that Jasmina has made recently. The stargets flagged by the algorithm, the diamond doesn't. If it's similar, machine learning ignores it. If it's too different, the transaction might not go through, as we saw happen to Jasmina.

So there you have it. The underlying mechanics behind one of the most useful unsupervised machine learning algorithms[21] that help keep us safe every day rests in a centuries-old distance function calculation from an ancient Greek. Thanks, Pythagoras!

Jasmina's dinner in Dallas with her high-school friend Alyssa is mostly unremarkable. While Jasmina can't resist the smoked brisket (she is in Texas,

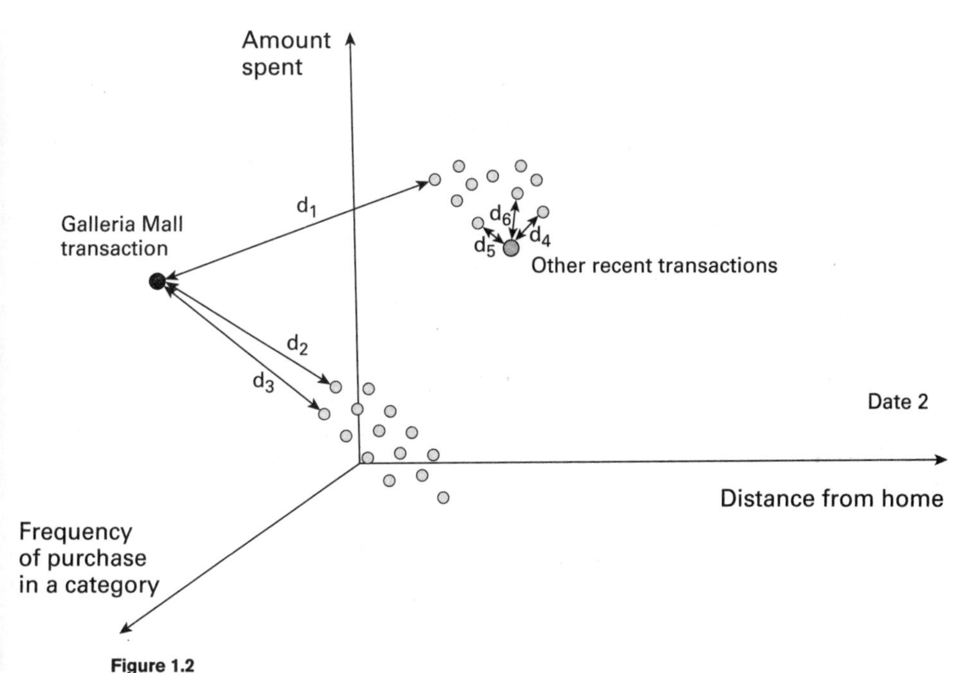

Figure 1.2

The gray dots reflect Jasmina's past transactions. The star is the new transaction being attempted at the Galleria Mall. We can see that the star-shaped point in comparison to the diamond-shaped point is far away from its three closest.

after all), Alyssa, formerly a voracious carnivore, has recently turned vegan. It turns out that a recent genetic analysis that she signed up for, when collated with other health markers, predicted that she is at high risk for a future occurrence of a certain cancer. It also turns out that this cancer is associated with her eating animal fat, at least according to her doctor.

This idea of "prediction" takes us into the territory of *supervised machine learning*.

In contrast to the banks raising red flags for transactions that fall outside of normal purchase patterns—the form of unsupervised machine learning we explored earlier—predicting future occurrence of a disease has a specific goal that it needs to achieve. In other words, the machine's process of "learning" from the data is going to be supervised or directed by its very specific goal of predicting future disease based on current demographic, genetic, and relevant medical markers. Such a specific goal to predict, also called an *outcome*, was absent in the anomaly detection setting where Jasmina's bank was looking for patterns in the data that seemed unusual.

Supervised machine learning-based prediction applications are everywhere. They take specific inputs, such as those we saw in the case of disease prediction for Alyssa, and use those to predict specific outcomes, for example, the likelihood of various life-altering diseases such as cancer. They are used by banks and financial institutions to model your risk of default before giving you a loan. The inputs for such a model could include a variety of inputs at the time of your loan application and could range from your creditworthiness, your debt-to income ratio, your zip code which could be a link to, say, census-level data about your surroundings, and other publicly available data about you (there is a lot out there). The output of such a model would be your likelihood of future default given what is known about you at the time of loan applications. If the likelihood is above a certain threshold, say 30 percent, the bank may decide against giving you the loan. We will address later in the book issues around a built-in bias against people of color who may live in formerly red-lined zip codes and have higher amounts of debt as a result of a previously (and still) stacked-against-them system.

How would the bank compute such a likelihood for your loan application? We will revisit our friend Pythagoras and his concept of distance between two points in a hyperspace. This is shown again in our now familiar three-dimensional visualization.

The bank has given thousands of loans in the last five years, and a small fraction of these defaulted. A simple and intuitive approach[22] to determine the risk of default would be to use the exact input data the applicant filled out when applying for the loan and find, say, the ten people most similar to that application from the historical loan data. Say only one of these ten defaulted. Using this, the bank would assess the applicant's likelihood of default as being one in ten, or ten percent, a low value, and give them the loan. On the other hand, if five out of the ten people most similar to the applicant when they applied for the loan defaulted, as is the case in figure 1.3, the bank would consider the risk to be 50 percent and would probably reject the loan application. Notice, that in contrast of figure 1.2 (unsupervised learning), which only had gray dots, in figure 1.3, we have two classes (defaulters and

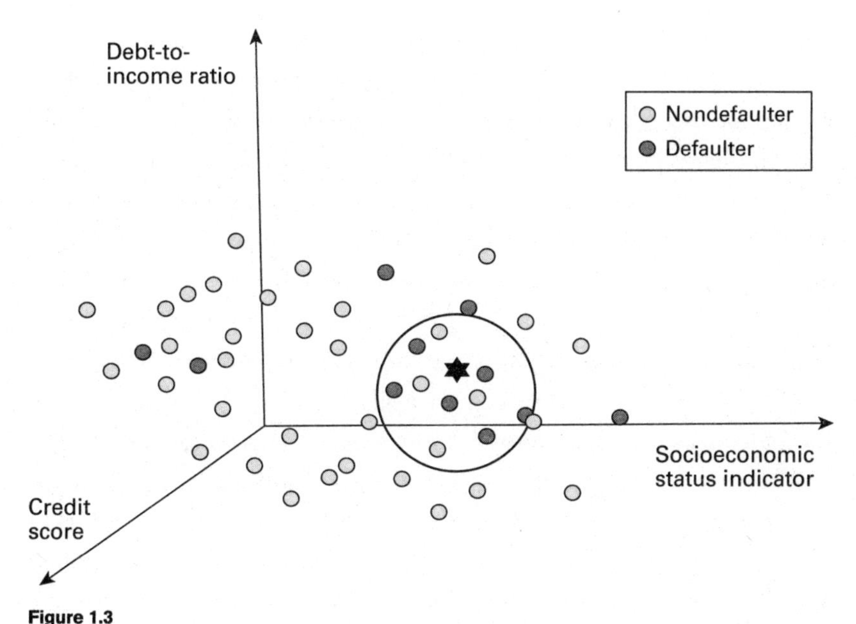

Figure 1.3

The black dots reflect prior loans that did not default. The hollow circles are those that defaulted. The star is the candidate's new loan applicant, and the circle around the star bounds the ten nearest neighbors, five of which corresponded to a default and four (almost five) of which did not.

nondefaulters) represented by black-filled dots and hollow dots (supervised learning, guided by the desired outcome). The job of a supervised machine learning algorithm is to use math and computation to find out what separates the two classes. In general, we can have multiple outcome classes. An insurer we worked with used to classify people into six different risk buckets corresponding to six outcome classes. In other cases, the outcome could just be a number, such as someone's cholesterol level.

Using the same idea of machine learning from past examples, Alyssa's medical provider was able to look at millions of past patients' records and compare them to her demographic, genetic, and medical markers and score her likelihood of risk from, say, heart disease, diabetes, and cancer as 0.01, 0.005, and 0.2 (1 percent, one-half percent, and 20 percent) respectively. While there are numerous models used to make such predictions and the scientific process of building and evaluating such models is robust, again, those are not the subject of this book. Instead, we hope that you now understand how supervised machine-learning models play a huge part in our daily lives. As you will see in subsequent chapters of this book, these models are used to screen resumes for hiring, to recommend movies and date candidates and products, to influence who you connect with on social platforms, to determine what specific nudges you get to exercise, to establish what ads you see on digital media, and of course, to predict machine failure of the most important machine: the human body.

Just as Alyssa is rightfully taking care of her diet after being forewarned of her risk for a certain type of cancer, one of us (Ravi Bapna) is also extra watchful about his cholesterol level, particularly, the so-called bad cholesterol (he remembers it as "L for lousy"), LDL, which his doctor said needs to be lower or he's at a greater risk for heart disease. The risk for Ravi is largely hereditary but also of great concern, since Ravi has lost several members of his family to heart disease. To monitor this risk and keep his LDL at a healthy level, he recently added an Apple Watch to his toolkit. This particular watch can take echocardiogram (ECG) data based on his heartbeat and rhythm and predict the onset of heart disease, which will give Ravi an early warning. He has also made other changes to his diet and exercise routine, the details of which can also be tracked by the Apple Watch.

To solve harder problems such as diagnosing heart disease using more complex data from ECGs, we need to introduce you to *deep learning*, a vibrant branch of AI that helps us make sense of the rich sensory data that we see in images, voice recordings, videos, and text written in various languages. Think of these as supervised and unsupervised machine-learning models that can handle complex data (complex for machines but intuitive for humans) in their raw form (say, pixels from an image, or audio waves from a musical piece) beyond just numbers in tabular format. A great example is that of two Danish researchers who built an AI solution using a deep convolutional neural network (a particular type of deep learning that excels in handling image data) to read 12-lead ECGs and diagnose heart attacks.[23] Today, deep learning models predict and complete the remainder of your phrase as you start searching on Google or type an email in Gmail. These models power Siri, Alexa, and other AI aunties that our kids now grow up with. They are the engines behind the promise of autonomous vehicles, as well as Generative

AI, which consists of a set of applications that generate human-readable language, remix new patents from existing patents[24] in an intelligent way, brand artwork, and generate music, among other things. At the time of writing there is much excitement around the release of OpenAI's GPT-4 (Generative Pre-trained Transformer 4), a large language model (LLM) that contextualizes the world's written text to predict the next word or phrase or sentence in response to a prompt.[25] We will present more on the working of LLMs in the context of education in chapter 5. After all, if students can rely on ChatGPT for all their answers for, say, AP exams, education as we know it is very vulnerable to be disrupted by Generative AI.

Over brisket and cauliflower tacos and after a few passionfruit mezcal margaritas, Jasmina and Alyssa settle into a lively argument over whether it's worth spending a large fraction of Alyssa's monthly salary for her to buy a new DSLR (digital single-lens reflex) camera. Alyssa is a property manager for multiple Airbnbs owned by an investor. Her compensation is tied, in part, to the amount of revenue the listings make per month. Alyssa has a hunch that improving the quality of the images she uses to list the various Airbnb properties should allow her to increase the property's price without reducing demand. Her boss is unconvinced, so if she wants to buy the expensive camera, she must use her own money.

Fortune favors the brave—especially if there is hard science lurking in the background, such as we find in the work by researchers from Harvard, Carnegie Mellon, and Boston Universities who used deep learning on images from 7,423 Airbnb properties over sixteen months and found that high-quality photos are associated with an additional $2,500 per year in revenue on Airbnb listings.[26] They took the raw pixels of these images as inputs and passed them through a series of deep learning layers that iteratively learned (from examples) features of these images such as their composition and aesthetic quality. The researchers were then able to associate improvements in these features with higher revenues. So Alyssa's intuition was spot on. If she buys the expensive camera, the increase in revenues from the listings she manages and the subsequent increase in her earnings will cover the cost of the camera in a few months.

Research like this is tricky, but beneficial. Despite the quality of the work and the interesting nature of the findings, the researchers had to persevere to publish their work in a top peer-reviewed journal. They needed to convince the reviewers that there were no lurking, nasty, common-cause variables such as the quality of the house, which would both yield good photos and high prices. Or that better photos were not in fact driving higher prices, but rather, higher prices allowed hosts to invest more in professional photography and other listing improvements—the nail of so-called reverse causality that kills many a paper. We will come back to the importance of causal analytics later in this chapter but consider this a sneak peek.

Alyssa does what most people do when they consider buying something. She enters "top DSLR cameras" into the Google search box and gets some targeted ads, some links to Instagram, YouTube and TikTok influencers, and a few relevant blog posts. But it's getting late in the evening and she has multiple properties to turn around over the next few days. She needs to ensure that guests write reviews of their stays and did not leave behind messes, and that housekeepers keep to their schedules. Before she realizes it, the weekend is upon her.

Over Sunday morning coffee, she checks into TikTok and Instagram, watches a few reels, and an hour flies by. A lot of the content she sees fits within what she expects. She loves picking up a few new styles and discovers a charming behavioral psychologist who gives life advice. She also encounters a paid influencer who gives tips on taking great iPhone photos. She watches that reel a few times, then moves on with her day, but not before up following this influencer. Later in the evening she navigates back, via another Google search query, to a review page titled "Top 10 DSLR Cameras."

Behind Alyssa's actions lies another type of machine learning (so far we have referred to unsupervised, supervised, and deep) that is working hard behind the scenes, called *reinforcement learning*. Its job is to pick up the data trails (e.g., how much time she spent watching a given reel) that Alyssa creates while multihoming across multiple digital platforms such as YouTube, TikTok, LinkedIn, Twitter (X), Amazon, Facebook, Netflix, and Instagram (among many others), understand their context, then give her a healthy mix of content and, where relevant, ads that will most likely appeal to her based on her history and interests. *Think of reinforcement learning as balance between exploration and exploitation.* The digital content and advertising ecosystem keep Alyssa engaged with content relevant to styling, whether for herself or for the properties she manages, vegan recipes, and the usual dance moves that most people her age love. These are her known preferences that the platform will "exploit" by showing her more of the same. But if it only does this, it fails to pick up on a new behavior or interest that may initially be a weak and perhaps noisy signal. For instance, she is interested in DSLRs but spent significant time on the iPhone photography influencer's reel. This is where reinforcement learning algorithms will "explore" a new category of content or advertisement for Alyssa. In fact, it was a display ad on Amazon that took an exploratory shot at showing a DSLR advertisement to Alyssa while she was catching up on her news that afternoon. Alyssa did not click on that ad, but research shows that such display ads tend to stick in people's memories and it probably increased the likelihood of her explicitly searching for "Top 10 DSLR Cameras" later that evening.

As we mentioned, reinforcement learning algorithms balance exploration with exploitation. They are used in a wide variety of settings where sequential decisions (what's the next best reel or ad to show to Alyssa) have to be made, or in cases where you do not have past data to apply supervised or unsupervised learning methods. These algorithms are showing promise in areas such as resume screening—a key time-consuming activity at the top of the hiring funnel—where traditional supervised learning methods have been criticized for perpetuating or even worsening human biases. For instance, Amazon was forced to shut downits AI-based resume-screening process once it was discovered that their new recruiting engine did not like women.[27] In contrast, other research finds that when companies use a reinforcement-learning approach that not only exploits past knowledge of what consists of good job candidates but also systematically explores newer types of candidates—for example, from minority groups—it is able to increase diversity while not sacrificing the quality of applicants in the hiring process.[28] Given the importance of fairness in algorithmic decision-making, as well as the generality of an explore-exploit strategy to make human life more interesting, we expect reinforcement learning to occupy a bigger role moving forward.

Alyssa's journey culminated in her getting a 25-percent-off coupon in her email from Best Buy, which she used to purchase a Nikon D7500 at a much lower price than her expected $3,000, especially after the AI-targeted discount. Through looking at Alyssa and Jasmina's week we were able to see the four types of machine learning at work—unsupervised, supervised, deep, and reinforcement—augmenting human abilities, reducing day-to-day frictions and generally making life more interesting. They make up a significant portion of the House of AI. Let's open the front door and explore what else constitutes the House of AI.

We conceptualized the House of AI (figure 1.4) as a framework that integrates the four types of machine learning we have discussed so far with the importance of a foundation of data engineering. It also encapsulates the causal and prescriptive analytics pillars, the emerging importance of Generative AI, an articulation of the types of decisions we make, and the critical role of ethical and fair use of well-translated analytics to benefit society.

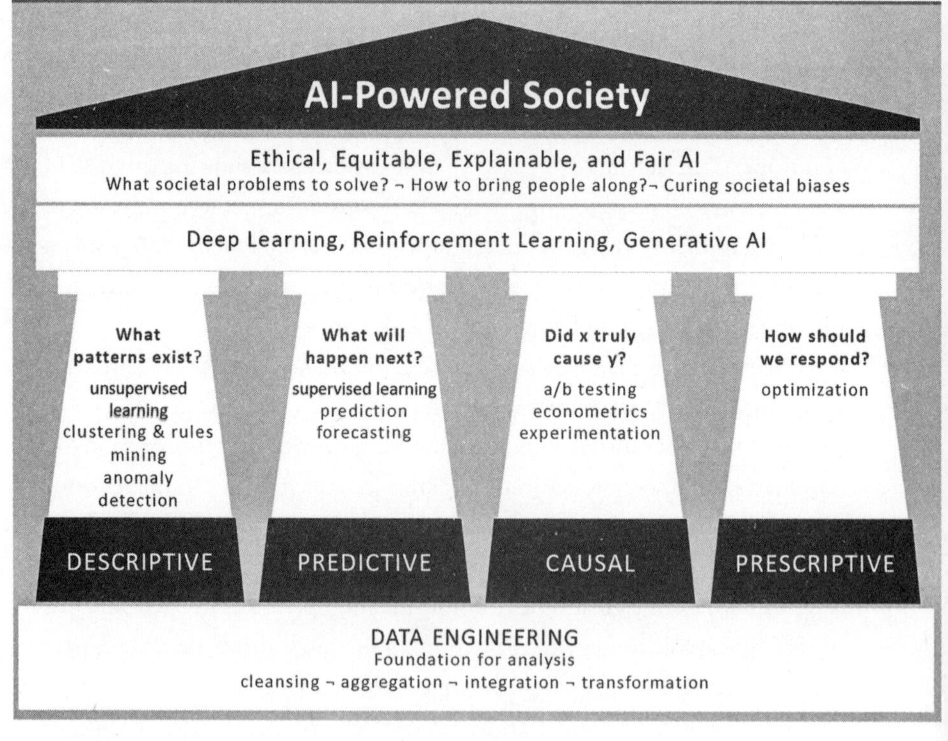

Figure 1.4
Multiple dimensions working together make up the House of AI.

Let's start with the foundation of data engineering. Our experience suggests that when constructing an AI project, close to 70 percent of the engineers' time is spent in first ideating what data to bring to the table for the four types of machine learning we discussed, and then cleaning, aggregating, integrating, and transforming them into something useful for further analytics. (As an aside, if you are looking for a career that is not going to go out of fashion anytime soon, look no further than data engineer.) Our house stands on the four pillars of data analytics: descriptive, predictive, causal, and prescriptive.

Starting with the low-hanging fruit, let's look at the first pillar, descriptive analytics, which relies on unsupervised machine learning. Descriptive analytics augments human intelligence by exploring and identifying interesting patterns in hyperdimensional data, something that we humans are not great at. (It *describes* these patterns.) We usually can't visualize beyond three dimensions. An example was the case of Jasmina's bank using anomaly detection to keep her safe from malicious actors misusing her credit card. Remember, instead of examining just three data points, the machine learning looked at thousands of data points. We also use descriptive analytics to group large amounts of data into meaningful and homogenous clusters (using an unsupervised machine learning algorithm called *clustering*) and to find patterns or co-occurrences of events (using, say, *association rule mining*).

Next, we have the second pillar, predictive analytics, which asks, "What will happen next?" (It *predicts*.) Will a consumer default on a loan? Will an employee churn? Predictive analytics overcomes yet another human deficit, our inability to explain or articulate how we make decisions. The Hungarian mathematician Polyani captured this concept in the statement "We know more than we can tell." So instead of asking tens or hundreds of loan managers how they decide the risk level of a loan, we use millions of rows of data from the bank's records of loans made in the past, some of which defaulted. Throw the scientific data-mining process of supervised learning at such data, let it learn from those millions of past examples, and out will emerge a best-in-class model that will be able to compute the risk of a new loan application.

Causal analytics, the third pillar, asks, "Does *X cause Y*?" It's a challenge for most decision-makers to step away from the data presented to them and ask for a counterfactual, using "what if" questions to consider alternate scenarios. Often the data is presented in the form of a report or a pretty visualization that may "quantify" the effectiveness of a new feature in the product, or, as was the case in a recent project of ours, the efficacy of a new mobile app channel that the brand introduced.[29] A well-meaning executive who funded an online dating app's development was pleased to see that engagement levels of customers using the mobile app was up compared to when the interactions were only web-based. Further, a new proposal was at his desk for additional app features that supposedly quantified the return on investment (ROI) from the app based on the value of the increase in engagement and the cost of developing the app. They key question that not enough executives ask, but *should* ask before believing ROI calculations such as these, is whether all the increase in engagement is a result of the app. This is the art of counterfactual thinking. Suppose you had a time machine and you could rewind time and not launch the app. Would you have seen the same increase in engagement? Could there be other factors that might possibly be driving the observed change? Often, people find that seasonality or promotions that others in the company might be running affect the engagement. And even bigger challenges arise when unimaginable and therefore unobservable factors (say, that sunny weather increases people's optimism and they are more likely to download the app and more likely to engage more) might be driving the outcome that was attributed to the intervention. What is the true cause?

Finally, the fourth pillar, prescriptive analytics, often combines many elements of the earlier three pillars and helps us model decisions that are situated in the context of organizational constraints, by asking "How should we respond?" (It *prescribes* or recommends actions.) Say a bank giving a loan wants to ensure that men and women have an equal true positive rate (that men and women who are truly eligible for loans are classified by the model as such) of being predicted as eligible (non-defaulting). There is no guarantee if we just use the predictive pillars that it will ensure such fairness, but a prescriptive approach would add this requirement (equal true positive rate for men and women or across races) as a constraint and optimize the decisions from the model's predictions to ensure fairness. Humans, again, are not naturally born with the gene that can optimize various decisions in the presence of constraints, nor can we operate without inherent biases. Machines, calibrated appropriately, can. Even more reason to augment human intelligence with AI.

In earlier research Ravi and his co-authors had laid out the House of Analytics as a framework that had data engineering as the foundation, the four pillars or analytics approaches (descriptive, predictive, causal, and prescriptive), and the importance of analytics translation.[30] In this book we have augmented the original House of Analytics with two new floors to propose the *House of AI*. Floor two belongs to the advanced concepts of deep learning, reinforcement learning, and Generative AI that are very much a part of our society today, which we will discuss further as we work our way through the book. The top floor of the House of AI is comprised of a set of norms, values, and practices that not only enable institutions to adopt AI—by making it explainable and translatable for laypeople to understand—but also to do so fairly and ethically.

If you have read this far, you are ahead of 95 percent of the world's population in your understanding of the most advanced general-purpose technology powering the fourth industrial revolution that is so integrated with our daily lives. In the chapters that follow we will take you into deeper into how AI can improve aspects of our lives that we all care about: finding love, fostering human connections, taking care of your health and wellness, bringing education to a new level, and ensuring financial savvy. Buckle up for this exciting journey: we hope you enjoy the ride.

TAKEAWAYS

- AI is a general-purpose technology powering the fourth industrial revolution. Like any significant technology, it has both pros and cons. So far, the narrative has been on the downside. This book aims to balance the discourse and encourage us to make AI work for business and society.
- AI is already having a massive positive impact on improving our everyday lives, including making society safer by reducing a variety of risks. Remember the example of the woman in Germany who had a seizure

and her Android phone called for help? Remember how the FCC is saving tens of thousands of lives per year? In the forthcoming chapters we cover various examples of how AI is having an effect on health, wellness, relationships, education, work, and our homes.

- We have proposed a framework called the House of AI to help laypeople become informed stakeholders in the discourse around AI. This book will give us an intuitive understanding and the agency that comes along with it, about what AI is, how it works through the lens of our everyday life, and how we can be an empowered citizenry in shaping it for our benefit.

2 FINDING LOVE

For most people, Jasmina included, one of the primary goals of adulthood is finding love. Obviously, there is no universal definition for "love," romantic or otherwise, and human ideas about love have been shaped by a dizzying array of factors—biological, historical, cultural, religious, economic, and sociological to name a few. In addition, individuals define love through the lens of their own desires, opinions, needs, personalities, emotions, and so on. Lucky for us, we don't aim to define love in this chapter; our interest lies in the *finding* of love—specifically, for the purposes of this chapter, romantic love.

Finding love is an essential part of the human experience. Countless stories, songs, films, works of art, and books and plays revolve around the complications that come with the search, the joys in the discovery, and the pain of loss. These books and movies don't just fall within the genres of romance novels and rom-com movies; it's rare to find an intricate storyline without some exploration of love. From *Scarborough Fair* to *I Will Always Love You*, from *Sense and Sensibility* to *Bridget Jones's Diary*, from *I Love Lucy* to *Sex and the City*, from *Romeo and Juliet* to *Hamilton*, and from *Casablanca* to *When Harry Met Sally*, the quest to find love permeates cultures across time and place. As a practical matter, "finding love" can be described in terms of the tangible, observable things people *do* to find romantic companionship.

An initial list of activities comes easily to mind. First, there are the *search* activities people undertake to meet potential partners. Attending social events, joining a club or interest group, asking others to make introductions, and going out to bars or night clubs are standard parts of most people's search. Next, *exploring or assessing* a potential relationship consists of dating, spending time together, conversing, and other efforts to determine compatibility and fit. Progressing along the continuum, humans *form romantic partnerships*. For many, that includes activities associated with *commitment*: cohabitation, sharing resources and responsibilities, marriage (whether defined in legal, religious, or other terms), having children, and caring for one another over time.

Not every person engages in these activities in the same way or with the same frequency, but no matter who you are and what you seek in a mate, the journey begins with meeting prospects and making connections with potential matches.

Connecting with potential matches usually means having to navigate a dating landscape filled with obstacles. This has always been the case, but unlike in previous generations, those coming of age during the last thirty years increasingly use technology to navigate the dating market. Internet connectivity, Web 2.0, social media, and smartphone apps powered by AI are just some of the technologies that now play roles in how people search for mates, assess potential relationships, and form romantic partnerships. Use of AI-powered online dating to find love has grown significantly over the years. In fact, of couples who met in 2017 (the most recent year for which data is available), *nearly 40 percent met online*, far higher than any other method of meeting a new partner.

Statistics like this have led many to wonder, "Just what kind of effect is AI-enabled online dating having on the human quest for love?" This is the focus of the remainder of this chapter.

Chances are good that you and many people you know can share an anecdote on the topic. After all, Internet dating is as old as the Web itself (the first dating website, Match.com, launched in April 1995[1]). Online chatrooms, instant messages, and dating websites and apps followed. If you were to ask people at a dinner party to share their experiences in this realm, most of the attendees would be able to share a story or two, some positive and some negative.

Jasmina tells the story of how her cousin met his significant other online and they've been together for ten years, but Alyssa can't help but think of how her email inbox was filled with unwanted graphic images within days of signing up for an online dating app. Other people's stories might be truly chilling ("My sister was stalked by someone she met online").

Given such a range of experiences, is AI-powered online dating good or bad when it comes to finding love? And what exactly does AI *do* in the various stages of finding love?

As we saw when Jasmina was protected against fraudulent credit card transactions by her bank's use of anomaly detection, first and foremost, modern AI technologies work hard at keeping us safe in online dating. This is often overlooked, but in many ways is as important as other aspects of AI that are more talked about, such as we saw in chapter 1 when Jasmina received personalized date recommendations based on multidimensional similarity. Even something as taken for granted as AI-assisted spelling and, increasingly, grammar correction can be life altering in the romantic context. As Grammarly CEO Max Lytvyn explains, "People use quality of writing as an indication of work ethic,"[2] and research by his company found that just two spelling errors on an online profile reduce the chances of receiving a response from a potential date by a full 14 percent.[3]

Let's unpack how AI keeps us safe first. Much in the same way Gmail encourages its users to tag emails as spam or phishing and then uses these "labeled" data to detect future spam or scams, AI plays a key role in detecting bots, fake accounts, and lewd images in online dating. Modern apps deploy deep learning models to detect and remove lewd images, and even detect offensive text if it's predictive of future harassment. Tinder, for instance, asks, "Does this bother you?" as a way to generate labels to then build supervised learning models to detect future use of offensive text. It uses this model to ask, "Are you sure?" in real-time if it detects a potentially abusive message. We tenured academics who tend to (and arguably are paid to) speak our minds would pay a lot for a chip in our brain that would occasionally apply the "Are you sure?" brake in a variety of life contexts. But that's a subject for another book.

How AI deals with textual data to match people or detect offensive speech may not be obvious, so let's lift the hood and take a brief excursion into the world of *natural language processing (NLP)*. Many of the ideas we develop here will also apply to other high-dimensional data such as images,

```
I'm into Mezcal margaritas, good hikes, and Revisionist History
(the podcast).
I love that where I live we arguably have the best Thai food
west of the Mississippi.
I'm most likely to initiate poker night but least likely to go
watch sports.
```

Figure 2.1
Data engineers convert text into numbers for subsequent use by AI algorithms.

audio, and video. We start with a sneak peek into a portion of Jasmina's OkCupid profile (see figure 2.1).

AI data engineers take such textual information and perform the so-called "text-to-numbers" trick to make textual data ready for unsupervised or supervised learning models. Then they scrub it to remove stopwords such as "a, an, the, from etc.," convert everything to lowercase (computers treat capitalized and uncapitalized letters differently), distill words to their roots (a process called stemming) so that fishes, fishing, fishery would all become *fish*, and delete very low-frequency and very high-frequency words. They do so for all the profiles on OkCupid that would be reasonable candidate matches for Jasmina. For Jasmina, our text-to-numbers trick may result in a one-row table with columns and a row of numbers as shown in tables 2.1.

Notice that "margarita" has become "margarit" after stemming. The same process applied to 2,000 other candidate matches for Jasmina will widen the table significantly (more "features" such as "learn" and "potter" from other users will appear) and introduce many zeros or blanks in Jasmina's row (see table 2.2).

At this point, this is still a crude cataloging of whether a particular profile contains a certain word or not. It does not, for instance, recognize that mezcal and tequila are close cousins when it comes to spirits or that there may be closer proximity between users who like other types of Asian food and Thai food in this region.

Table 2.1
Numeric representation (partial display) of Jasmina's OkCupid profile

mezcal	margarit	hike	Revision	history	argue	best	thai	. . .	. . .
1	1	1	1	1	1	1	1		

Table 2.2
Numeric representation (partial display) of 2000 users' OkCupid profiles

							Column 3,000 ↓	
Users	**mezcal**	**learn**	**potter**	**tequila**	**takeout**	**margarit**	**india**	**. . .**
1 (Jasmina)	1				1	1		1
76				1		1		
. . .								
2,000			1		1		1	

To bring meaning to these dimensions, to recognize that certain concepts are close to certain other dimensions but far from others, AI data scientists perform another *embedding* trick that imagines that these texts are being generated by some small set of latent dimensions. These are dimensions that might not be directly observed but can be *assumed* by AI. In our example, it could be that these dimensions map to choices of spirits, food, and physical activity. These dimensions themselves are discovered using deep learning, but the important things to note are that (1) they are hidden in the data, and (2) with enough data they can be discovered. This takes us now to where we started in chapter 1. Instead of explicit dimensions of income, height, and attractiveness, we have three latent dimensions that (roughly, upon examination after the fact) are associated with choices of spirits, food, and physical activity (see figure 2.2).

Based on our model, User 76 is closer to Jasmina than to Users 1756 and 2000 in a latent hyperspace consisting of the language and specific terms these individuals used in their profiles. The same embedding concept can also include pixel data from images. Together with traditional numeric attributes such as income, height and attractiveness, they give modern AI an excellent shot at finding a great date for Jasmina.

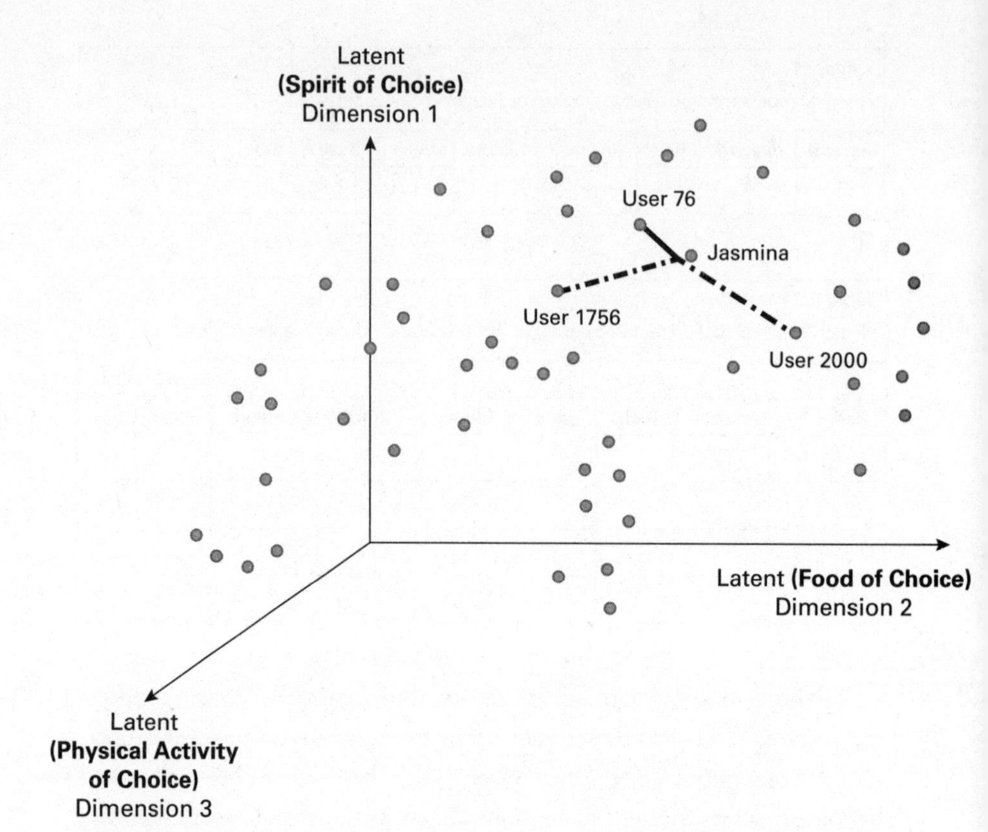

Figure 2.2
Embedding compactly represents Jasmina's 3,000-dimensional textual information in three latent dimensions.

We can now apply the *text-to-numbers-to-embedding* trick to comments (sentences, phrases, chats), and if we have users who have tagged certain words as offensive, we can apply everything we know from the supervised learning predictive pillar of chapter 1 to flag an "Are you sure?" warning in real-time.

THE EFFECT OF AI-POWERED ONLINE DATING

Returning to the question of the broader effect of online dating on both regular people's lives and society as a whole, Jasmina's and Alyssa's experiences and those of others like them, while potentially interesting, don't allow for clear conclusions or verifiable answers as to how AI has changed our search for love. The preceding stories and many others may represent the experiences of individuals, but which stories are examples of broader truths, and which are exceptions? As researchers our goal is to find answers supported by hard evidence.

It might be surprising to know that quite a bit of academic research on dating and technology already exists. In part, this is because digital technologies generate large quantities of data that can be studied. Real-world settings like bars have no built-in data-collection features, but online platforms and apps generate huge amounts of incredibly detailed data. These systems can record basics such as the number of clicks and page views, as well more advanced information such as how much time a user spends looking at another user's profile. Multiplied by millions of users, these systems generate massive amounts of rich data.

Researchers find such data sets valuable because they provide tangible signals of actual human behavior rather than simulated behavior from a lab or projected behavior from a questionnaire, which are often less true to the way humans behave in real life (people think they will behave a certain way, then do something different in the moment). In the case of finding love, these data represent proof of people's *actions*. Careful study of such data can yield new knowledge that expands what we humans collectively understand about the world. A note before we go any further: Concerns about personal information are certainly understandable in this context, however, we can assure you that academic research follows strict ethics standards and other protocols. (The studies we conduct and cite certainly do.) Researchers aren't interested in the behaviors of specific individuals; they're using anonymized data to investigate aggregate trends.

So how did online dating change from a novelty to the principal means for people to meet romantic partners? Researchers have heralded the rise of online dating as something of a revolution in matchmaking. One author says that finding a romantic partner "is one of the biggest problems that humans face," and that the advent of online dating was "one of the first times in human history there was some innovation."[4] Another states, "There have been two major transitions" in heterosexual mating "in the last four million years. The first was around 10,000 to 15,000 years ago, in the agricultural revolution, when we became less migratory and more settled. . . . And the second major transition is with the rise of the Internet."[5]

No wonder people flocked to dating sites: 10,000 years is a long time to wait for innovation.

From its inception, online dating has enticed users with a handful of appealing features, many of them marketed as keys to unlocking more successful matches. To begin with, computing technologies offer scale and speed, both of which lower search costs. As an article in the *American Sociological Review* notes, "Searching personal advertisements in the pre-Internet era meant thumbing through the newspaper classified section by hand. Print advertisements could only be examined one issue at a time. . . . In contrast to the inefficiencies of searching paper documents, online search makes the archive of old issues just as accessible as the current issue. Online, it is as easy to search across a million records as to search across a hundred."[6]

Scale and speed mean more choices, and more choices suggest better chances of finding a match. Even the most social of butterflies can't compete with the Internet's reach. As a recent study explains, "the sets of people connected to Tinder, Match, and eharmony are larger than the sets of people connected to one's mother or friend. Larger choice sets are valuable to everyone engaged in search."[7]

Anonymity is another important feature. Online dating platforms offer users the ability to view potential matches from a distance, mediated by the computer. That makes anonymity possible and offers a degree of discretion. In addition, users can screen potential matches before meeting in person, providing a measure of safety relative to meetings in bars or on blind dates set up by friends.[8] Additionally, because messaging in online dating is asynchronous, users can better present themselves positively and deliberately, particularly those who might suffer from social anxiety in offline settings.[9]

Finally, the power of AI-enabled online dating promises *better matches*, at least relative to pre-Internet matchmaking services (including those offered by friends, family, or coworkers).[10] For instance, eharmony, which launched in 2000, claims that its "unique Compatibility Matching System measures each potential couple on thirty-two dimensions of compatibility."[11] Match.com claims that its algorithm, codenamed "Synapse," builds on the mountain of data from the 75 million users it has had since it was founded, incorporating such insights as the finding that women are less likely to email with men who live far away, men who are older than they are, and men who are short. Other findings are more nuanced. Catholic women are especially unlikely to email a Hindu or atheist male. And while men are most particular about hair color, a woman's income is less important to them.[12]

Another way to view the appeal of online dating lies in how these features reduce certain *social frictions* individuals navigate within communities and society. As you likely know from experience or from having seen reporting related to online anonymity, people's behavior often changes when they know they cannot be identified. The results of such disinhibition can be substantial, as has been shown in a number of research studies.[13]

Certainly, one can argue that disinhibition is detrimental when it results in harassment in the form of trolling, bullying, stalking, and other undesirable behaviors, but when it comes to finding love, disinhibition through anonymity may offer something more positive. Because their activities are relatively shielded from societal pressures to conform to particular norms, people can search more freely online. For example, real-world social frictions may inhibit people from pursuing what might be considered taboo, such interracial or same-sex relationships,[14] but online environments and anonymous search features may reduce such stigma and facilitate matches that might not have happened in the real world.[15]

DIGITAL DATING: GOOD, BAD, OR INDIFFERENT?

Over its history, online dating has gone from a novelty to the mainstream in short order. Dating platforms have evolved along with computing technology to offer new and different ways for singles to find, meet, and connect with romantic partners. Almost thirty years into digital dating, what have we learned? What effects do big data, algorithms, artificial intelligence, and other technologies have on finding love? Are there enough positives to counter the cringe-worthy anecdotes?

As you likely suspect, the answers to these questions are more nuanced and more interesting than a simple "yes" or "no." The fact is that human relationships are complex and messy even at their best. Although various digital-dating platforms have promised better matches, it's difficult to define what "better" means in this context, much less prove whether the promise is fulfilled or not. *Better* is a slippery concept when it comes to matchmaking, and in the end, success is defined by what happens after people meet one another. What ultimately determines whether a couple finds the love or kind of relationship they hope for depends much more on the conversations, dates, activities, and bonds that are formed (or not) after they meet than on the specifics of the dating platform. Still, there is evidence that digital dating has eroded many traditional barriers, especially at the initial matching stage, which is progress.

THE EFFECTS OF ANONYMITY

As we previously mentioned, digital environments and the data they generate offer new opportunities to study various phenomena. As a perfect example, we can look at research on anonymity and online dating.

The ability to remain anonymous while screening potential dates is not practically possible in the physical world, but online, anonymity is not only possible, it's something social networks have incorporated by design for quite some time.

Dating platforms may advertise anonymity as an incentive to search for love online in order to shield users from social scrutiny. Users may agree this is an advantage, especially those who seek to escape certain social, cultural, and religious rules for what is acceptable and what is taboo. However, what about the costs or misunderstood effects of such a feature? On the surface, anonymity may seem all upside—greater privacy, personal safety, freedom from social pressures, etc. But are there drawbacks, too?

In an earlier study, Bapna and his coauthors addressed the question this way:

> On one hand, anonymity is likely to lower search costs and therefore lead to *disinhibition*. Users need not worry about how others interpret or perceive their visits, possibly even repeated visits that may otherwise be considered "stalking."

On the other hand, with respect to social frictions, anonymity may impact the matching process via *signaling* related mechanisms. In particular, by hiding the focal user's actions from others, anonymity may influence signaling protocols that are necessary to establish successful communication with a potential mate. Thus, broadly speaking, our research objective is to examine the net effect of disinhibition and signaling in online dating.[16]

This study included a large experiment conducted on a popular dating website. The researchers randomly selected 100,000 users and gave half of them the ability to browse other users' profiles anonymously, while the other 50,000 users did not have the ability to browse anonymously. Importantly, users with the anonymity feature turned on still had a visible profile on the platform and they could contact other users directly via messaging. The only difference was that users in the first group had the ability *browse* others' profiles anonymously.

The results of the study showed that the anonymity feature did indeed lead to disinhibition. Users with the anonymous browsing feature enabled *viewed a wider variety of profiles*, including those of same-sex users and members of other ethnic groups; however, the anonymous users *did not actually make more matches* than those without the ability to browse anonymously.

Obviously, a successful match is the goal. If anonymous browsing leads to fewer successful matches, despite users' disinhibition, what could be the reason?

The answer has to do with a different kind of social friction: how people show interest in one another without being "too forward" or coming on too strong. This, too, is governed by social norms, and women are subject to much more scrutiny in this regard. Traditional gender norms in heterosexual mating strongly discourage women from "making the first move" when dating.

In the *online* dating world, making the first move usually translates as sending a message to another user to start a conversation. In the dating site study we described, users in both groups had this ability. However, the users with the anonymous browsing featured turned *on* lost a different ability, namely, the ability to show interest in a more subtle, less direct way. Known as weak signaling, this is definitely important when it comes to achieving not just initial messaging, but also successful matches.

In the study, a weak signal is codified in a "who viewed you?" feature. (Similar features are built into many social platforms.) Essentially, if a user (let's call her Susan) visits another user's profile (we'll call him Ryan) but does not take the *explicit* step of sending Ryan a message, *the visit itself* may be interpreted by Ryan as a signal of interest. In visiting Ryan's profile, Susan leaves a weak signal (whether she means to or not).

If Susan's visit to Ryan's profile is visible in a "who viewed you?" list, then Ryan may be encouraged to message Susan, thus initiating a match. But that can only happen if Ryan sees that Susan visited. If a user like Susan is anonymous by choice or platform design, she loses the weak-signal functionality and decreases her chances of matching with other users. The study establishes that such weak signaling is causally linked to successful matching outcomes.

Furthermore, the ability to leave a weak signal is shown to be more important for women than men. Because women are traditionally discouraged from "making the first move," leaving a weak signal allows Susan to communicate interest without breaking a social norm. Susan did not make a *definitive, explicit* move, and yet her appearance on Ryan's "who viewed you?" list *implies* interest. Implying interest via a weak signal facilitates the match for both parties: Susan does not have to break the social norm and yet her weak signal serves as a kind of invitation for Ryan to initiate contact.

Another noteworthy aspect of these results is that the evidence of weak signaling's role would be practically impossible to detect and prove in the physical world. The closest real-world equivalent to the "who viewed you?" signal would likely be flirting behaviors. However, a smile, a lingering glance, laughter at someone's joke are much more ambiguous and open to interpretation. And good luck trying to study such behaviors in 100,000 people seeking dates at a bar.

Other research shows that the ability to leave a range of signals in the digital world significantly increases the number of matches for both women and men. In a twist on the "who viewed you" feature, another study examined the effect of a "who likes you" feature on another online dating site. The results showed that when women can see the identities of men who "liked" their profiles, they became more proactive and sent 7.4 percent more direct messages to other users and increased their matches by 14.4 percent.

The same feature benefits men seeking matches. Without the "who likes you" feature, men send about twice as many messages as women, but sending more messages doesn't automatically mean more matches. In contrast, men increased their matches by 11.5 percent when the "who liked you" feature was turned on.

This leads us to the next iteration in digital dating: apps that focus on matching likes with likes.

THE APPS: PHOTOS, GAMIFICATION, AND AUTHENTICATION CHANGE THE SCENE

The introduction of mobile apps for dating added some new features not found on legacy dating websites. Mobile apps provide convenience and allow users to check dating prospects throughout the day on their smartphones,[17] their push notifications encourage additional engagement, and the GPS technology in phones provides the ability to search for nearby matches.[18] These features helped drive the surge in online dating's popularity in the 2010s: Grindr launched in March 2009,[19] Tinder launched in September 2012, and Bumble launched December 2014.[20] Legacy dating websites also launched their own apps to keep pace.[21]

Tinder in particular was responsible for a large share of the growth during the 2010s. In May 2013, OkCupid was the largest online dating service, with one source indicating a 55 percent market share at that time. Match .com followed with approximately an 18 percent share. Four months later, Tinder was the largest online dating service, with a more than 50 percent market share; it reached an 80 percent share by early 2014, not even two years after its launch.[22]

Clearly, Tinder's entry into the market dramatically changed the landscape, but why? Tinder was a late entry compared to the legacy sites that had first-mover advantage and large user bases. Those factors should have worked against Tinder. Research points to several reasons for Tinder's rapid ascent.

First, unlike the legacy sites touting complex matchmaking algorithms based on lengthy questionnaires, Tinder used simple photo-based profiles, with text limited to basic information and a 240-character biography. Despite seeming to provide a wealth of information, the legacy sites' approach often provided little that was truly relevant to users. As one group of researchers explained:

> Most online dating sites use a "shopping" interface like that used by other commercial sites, in which people are classified much like any commodity, by different searchable attributes (e.g., height, weight, income), which can be filtered in any way the shopper desires. . . . Because determining whether or not one likes someone romantically requires subjective knowledge about *experiential* attributes such as rapport or sense of humor, it is perhaps not surprising that online daters might be disappointed when they are forced to screen potential partners using objective *searchable* attributes such as income and religion.[23]

Tinder's interface design focused on experiential attributes of appearance and attraction. As one case study explained, "Its fast, frictionless matching process allowed users to quickly express positive interest in others by swiping right (like) or negative interest by swiping left (pass) based on user photos."[24] As we have discussed, when users can see who "likes" them or even views them, matching increases.

A second factor in Tinder's growth was that it appealed to young adults. Between 2013 and 2019, the share of 18- to 24-year-olds who reported using mobile dating apps increased from 5 percent to 22 percent.[25] Tinder was first promoted on college campuses, and the app charges more to users over the age of 30.[26] Traditional dating sites such as Match.com and eharmony catered to older demographics, so prior to the appearance of Tinder, survey data found that "about 23 percent of people below [the age of] 40 started a relationship through the Internet, whereas about 36 percent of people 40 and above found their current partner online."[27] Other studies of the pre-app period called it a "puzzle" that "young heterosexual adults, who are presumably among the most technologically savvy people in society, are among the least likely to meet partners online."[28] Tinder's success was therefore an example of "disruption" in online dating since it identified and appealed to a previously overlooked group of users.[29]

Tinder also incorporated elements of "gaming" to the quest to find partners, namely the iconic "swiping" element.[30] On the Tinder app, users rapidly view many profiles and swipe right or swipe left to indicate interest. Matching unlocks rewards in the form of messaging, and successful matchers are more highly ranked in subsequent searches by other users.[31]

Gamification is a recognized element of success in mobile apps across many genres and has been incorporated into other popular apps including Reddit and Snapchat.[32] One study of app gamification states that apps like Tinder "opt for simplified, visualized user interfaces with little information, integrating swiping, pop-up notifications and other game-like elements. Thereby, they differentiate themselves from online dating sites, the operating logic of which is based on hyper-cognized compatibility questionnaires and matching algorithms."[33]

Tinder users actually see the app as a game, at least in part. A summary of user reviews on the app finds that "Tinder's users described their experience with the app using words related to 'fun' more than twice (2.5 times) as often as competing dating applications."[34] A set of twenty-one ethnographic interviews with Tinder users in 2014 found that "when initially asked why they used Tinder, all but two of the interviewees described their use as entertainment or an ego-boost. Erwin, thirty-four years old, identified wholeheartedly with the notion that Tinder was just for entertainment: 'For me it's more like a game.' To illustrate, he pointed to the games folder on his iPhone. 'See, the app is right here, right next to Candy Crush.'"[35]

Tinder's creators have publicly stated that they purposely developed the app with gaming in mind. Sean Rad stated, "We always saw Tinder, the interface, as a game. What you're doing, the motion, the reaction."[36] Jonathan Badeen said, "Because we've wanted to keep things lighthearted we took inspiration from games."[37] Badeen has also stated that he was inspired to develop the swiping element of Tinder based on psychology experiments by B. F. Skinner on pigeons, in which the birds are conditioned to believe that their pecking prompts the delivery of food, when in fact the food is delivered randomly.[38]

Scott Hurff, formerly the product manager and lead designer at Tinder, writes that "Tinder is one of the most addictive experiences, period, with an estimated 50 million people using the product to meet up, date, marry, and more. Since the swiping gesture is so easy and natural to perform, it encourages quick feedback on someone's profile. But that's not what keeps people swiping. It's the gambling-like reward, that dopamine rush of the 'It's a Match' screen that descends upon you without your knowledge."[39]

A final factor that differentiated Tinder from other dating sites and apps was that users are required to link to their Facebook pages, thus providing a kind of authentication for Tinder's profiles, despite their sparseness. Online dating has long raised concerns about safety, both online and offline. For instance, according to a 2019 survey of online dating, 48 percent of women and 27 percent of men had someone continue to contact them online after saying they were not interested; 46 percent of women and 26 percent of men received a sexually explicit message or images they did not ask for; and 11 percent of women and 6 percent of men had suffered a threat of physical harm in online dating.[40]

By linking Tinder profiles to actual people with histories of photos and other content on Facebook, individuals with potentially malicious intent can be better screened out. As one author explains,

> Tinder's welcome screen presents a large blue button prompting users to "Log in with Facebook." It is not possible to use the app without taking this step. Since dating through Tinder involves evaluating others' self-presentations, exchanging geolocation information, and meeting unknown others—presenting the potential for misrepresentation and threats to personal safety—Facebook is Tinder's safeguard against the uncertainty this may cause. . . . While other dating apps have since integrated with Facebook, Tinder's approach was novel upon its launch in 2012. Concerns over information accuracy and security are appeased by constructing a profile form a Facebook user's first name, age, recent photos, and gender while storing likes and friend lists to display mutual friends and interests in browsing mode.[41]

Similarly, Tinder does not allow users to post photos directly from their phones; images are drawn from existing Facebook or Instagram pages.[42]

BUMBLE: MAKING THE FIRST MOVE

One of the people behind Tinder's early success was cofounder Whitney Wolfe Herd. But Wolfe Herd is perhaps better known for the company she founded next: Bumble. As CEO, Wolfe Herd's leadership of Bumble has earned her many distinctions, including the monikers of "world's youngest self-made female billionaire" and "youngest woman ever to take a company public."[43]

As of 2021, Wolfe Herd's Bumble dating app was worth $14 billion and had forty-two million active users each month. But these impressive numbers only tell part of the story. What's most notable to us about Bumble's success is the app's unique approach to designing a digital dating platform.

Wolfe Herd and her colleagues at Bumble have created an online dating experience centered on women and women's priorities as users of the platform. Wolfe Herd has said in interviews that gender role assignments impacted her from a young age.[44] Bumble's design puts women firmly in control. Matches between users only proceed when women "make the first move." That is, women must initiate contact with a match. "Make the first move" is also Bumble's tagline, which seems appropriate, given that the app turns that particular social friction on its head.

Bumble's design reflects academic research on the topic, suggesting that Bumble users are better positioned to find a match versus other platforms. In addition, Wolfe Herd and her team have taken on many of the more unsavory elements of online dating. For example, Bumble has software to proactively blur lewd images. Combined with the feature that allows only women to start conversations with other users, such engineering dramatically cuts down on unwanted advances. Wolfe Herd has even successfully lobbied lawmakers in Texas, where Bumble is headquartered, to make sending unsolicited nude images a crime.

Efforts to engineer a safer, more pleasant online environment have earned Bumble praise. Calling Bumble "wholesome" or "the least toxic dating app" may not seem like obvious compliments, but then again, much of the Internet remains an unregulated Wild West of trolls.

Profiles of Wolfe Herd in *Time* and *GQ* have given Bumble favorable reviews. r. Where some platforms have an inconsistent, even laissez-faire approach to codes of conduct and behavior standards, Bumble embraces rules and guardrails. Wolfe Herd cites acceptable behavior in the physical world as Bumble's model for how the company thinks about what's acceptable on the platform and what is not. For example, indecent exposure is a crime in the physical world; therefore, the equivalent online behavior (sending unwanted photos) is off limits.

As the author of a *Time* profile wrote, "More than relationships, or friendships, or in-app purchases, Wolfe Herd is selling the feeling of power to the powerless, a sense of order in an online universe that so frequently seems lawless."[45]

DATING IN THE METAVERSE AND BEYOND

What's ahead for people looking for love online? As the Internet becomes ever more embedded in how people communicate, work, and socialize, we believe online dating and the resulting real-world quest for love will continue to evolve alongside the technologies humans use.

For instance, although still nascent in many respects, the metaverse, where users can interact with other users in a virtual reality (VR) space, may be the next dating frontier. Emerging interest in the metaverse on the part of Facebook founder Mark Zuckerberg and others has brought increasing attention to virtual reality, augmented reality, and related technologies. If these things all come together in the metaverse as Zuckerberg and others promise they will, humans will no doubt experiment with using the metaverse for all kinds of activities, from remote work and commerce to socializing and, yes, dating.

In fact, it's already begun. As of this writing, a team of collaborators has launched thedatingverse.com, a service offering "virtual reality date coaching." According to the website, "The combination of VR and date coaching is a game changer" and "VR practice makes IRL perfect."[46]

Along similar lines, the documentary *We Met in Virtual Reality* appeared on HBO's streaming service in mid-2022. Its description says, "The film reveals the growing power and intimacy of several relationships formed in the virtual world, many of which began during the COVID-19 lockdown, while so many in the physical world were facing intense isolation."[47]

At the same time, the human need to meet, look people in the eye, catch a sunset, sleep under the stars, and find emotional and physical connection will never go away.

TAKEAWAYS

- Can AI help us find love? Yes! AI-powered digital platforms are enabling more than 40 percent of new romantic relationships in the United States. AI-powered tools such as Grammarly are helping people refine their online profiles given that just two spelling errors on an online profile reduce the chances of receiving a response from a potential date by a full 14 percent. Gamification is a powerful driver of engagement in mobile apps across many genres and dating apps have incorporated it, as have other popular apps including Reddit and Snapchat.
- Modern AI technologies work hard at keeping us safe in the world of online dating. Deep learning models on Tinder can detect lewd images and even detect offensive text and hate speech, which are often predictors of future harassment.
- AI can richly represent the multidimensional profiles of individuals in a contextual space called *embedding*, which is the basis for superior matching algorithms. This results in increased diversity of romantic options and reduction of other social friction such as anxiety. The AI-powered digital ecosystem empowers women and gives them agency in making the first move.

3 FOSTERING HUMAN CONNECTIONS

Perhaps someone you know has a story like this: It's January of 2021, and thankfully, the pandemic is mostly behind us. Jasmina parks her car a few blocks up the street from her house. It's busier than usual for a Thursday evening. As she walks to her place, two women walking up the street see her and cross to the other side. They walk half a block on that side, then cross back once they are past Jasmina. It's hard for Jasmina to attribute this to social distancing. She is disturbed but has a busy evening ahead of her, so she lets go of the negative thoughts regarding being racially profiled in her own neighborhood. After dinner, she remembers that she needs to sell the extra blender taking up space in her tiny kitchen and logs onto Nextdoor, a hyperlocal social networking service for neighborhoods. Before she can post her blender for sale, the first post she sees after logging on enrages her. Her agitated neighbors are warning residents of "a shady black man in a hoodie just loitering," and they're asking for the police to be called. Jasmina's experience is not unique or isolated. For instance, other stories have reported how all the vital information about Black Lives Matter protests were drowned out by comments of "All Lives Matter" or tagged as riots on Nextdoor.[1] In the summer of 2020, this led to CEO Sarah Friar publicly apologizing and taking blame for not doing enough to moderate Nextdoor's volunteer moderators.[2]

Fast forward to March 30, 2022, and we were surprised to see Sarah Friar's picture on the "TIME100 Most Influential Companies of 2022" list for "Promoting kindness." Representatives at Time said:

> In response to criticism that it was failing to curb racist content, neighborhood social networking platform Nextdoor and CEO Sarah Friar last year launched

a system that scans posts for red flags and encourages users to reconsider before publishing something problematic. The feature builds upon Nextdoor's Kindness Reminder feature, which similarly nudges users if it detects negative language or other signs of a heated conversation. Nextdoor says people shown such prompts edit or withhold their content about a third of the time, making them a promising solution for any social media company struggling with toxicity.[3]

Digging a bit deeper, we found a new feature called "Constructive Conversations Reminder" on Nextdoor that is similar to what we saw in chapter 2 on AI detecting lewd images or nasty language on Tinder. Nextdoor uses AI and ML to predict and detect, in real-time, when a comments thread may become heated *before* a neighbor contributes.[4] When the AI model predicts a high risk of a thread getting nasty, the app intervenes with a reminder and some tips on how to write with a more empathetic tone. What really caught our attention was a statement that said, "More specifically, the ML model considers the full conversation rather than individual comments." Let's unpack the value addition and some key challenges of building an AI model that considers a full conversation rather than individual comments. This takes us a bit deeper into our excursion into natural language processing (NLP).

One of the earliest applications of NLP was taking a review of a product and classifying it as positive, neutral, or negative. This involves the text-to-numbers "trick" we discussed in the last chapter and is formulated as a supervised learning classification problem. Another popular NLP task called "name detection" allows machine learning to be more streamlined and personalized. Given a sentence or a paragraph, this AI task is to examine words and detect whether they correspond to a person's name or not. In the sentence that follows, the words in bold obviously correspond to a (fictional made-up) name:

José Ramos was not only a great dancer but also a fine president.

However, we need the AI to be smart enough to recognize that in the sentence below:

Don **José** distillery makes a fine rum.

José (in bold) does *not* correspond to someone's name, but a business. We need it to consider that, in language, the words that come later in a sentence can be predictive of what came before. This is where modern advances in deep learning models come into play that can detect bidirectional patterns in sequential data, often carrying memory and context from words said at the beginning of a conversation all the way to its end. For instance, AI can complete your sentences (you may have seen this on Google search or in your Gmail app) by predicting what the next logical word will be in a sentence such as:

The pitcher, filled with the most delicious lemonade, was always half full.

And learn that when AI sees:

The pitchers, filled with the most delicious lemonade,

It will complete the sentence as:

The pitchers, filled with the most delicious lemonade, were always half full.

The AI knows the connection between a word in the beginning of a paragraph and the end of it and adjusts later words accordingly. Such is the power of deep learning models such as bidirectional recurrent neural networks (BRNNs)[5] or long short-term memory (LSTM)[6] that are trained on enormous corpuses of text such as Wikipedia to learn our languages. Models such as these and their successor transformer models that power LLMs such as GPT-4 are the source of much excitement in the AI community. More on the inner working of such models in chapter 5.

BRINGING STRANGERS HOME

In January 2018, one of the authors, Ravi, and his wife, Sofia, became Airbnb hosts. The home they furnished, styled, and listed on Airbnb was not their house in Minneapolis, but a flat in Jaipur, India, a place they visited once or twice a year in order to spend time with Ravi's parents and other family.

The flat in Jaipur offered many benefits. Instead of staying in anonymous hotel rooms when they visited, they stayed at the flat. They had more space to spread out and their teenage daughter had a room of her own. They could entertain guests comfortably and even host friends overnight if the occasion called for it. Thanks in part to the flat, their visits to Jaipur felt more like coming home.

Like so many other Airbnb hosts, VRBO members, and car owners who sign up with companies like Getaround and Turo, Ravi and Sofia used technology to put one of their biggest assets to work for them.

A key aspect of getting the economics to work, especially in the early stages when they did not have many five-star reviews, was to set the correct price for the rental to drive demand. It turns out that this is more complicated than looking at the average price of other three-bedroom Airbnb flats in Jaipur. How would they account for the look and feel, the art, and the top-notch soft furnishings that they had invested in, not to mention the additional value of having a washer-dryer and dishwasher (rare for India where household staff is inexpensive) in the fully equipped kitchen? Determining optimal pricing in early stages versus in order to have their Airbnb steadily occupied was complicated.

This is where AI intervened in the form of Airbnb's *smart-pricing algorithm*. Think of this as a back-end tool for hosts—a supervised learning model—that predicts demand given a chosen price and a number of other factors. These factors range from the hard attributes of the property—for example, the number of beds, baths, the amenities, and the appliances—to the incremental value of the soft furnishings modeled through image analytics, to seasonality, competition, and how far in advance people book. Not only are these features considered in and of themselves, but the AI algorithms also consider their myriad interactions, such as the additional value of a "three-bedroom flat—with tribal art—opposite Central Park—with a dishwasher—two weeks prior to a major event such as the Jaipur Literature Festival (known to attract close to 400,000 visitors to the city)."[7] Users can use this tool and set a range of prices based on certain conditions, which Airbnb then further dynamically optimizes on a daily basis.

It helped tremendously that Airbnb's smart-pricing AI tool took care of the day-to-day optimization of pricing for the flat so that Ravi and Sofia could focus their energies on visiting with Ravi's parents while they were in India, and his parents could focus on meeting travelers from all over the world when they were not.[8] Recent research has shown that hosts who adopt this smart-pricing algorithm experience an 8.6 percent increase in daily revenue on Airbnb.[9] Interestingly, the researchers also found a stark racial divide in who benefits from the adoption of AI. Before Airbnb introduced the algorithm, white hosts earned $12.16 more in daily revenue than Black hosts (controlling for observed characteristics of the hosts, properties, and locations). When both white and Black hosts used smart pricing, the revenue gap between them decreased by 71.3 percent. However, Black hosts were significantly less likely than white hosts to actually adopt the algorithm. This may be because of a lack of trust in AI-based algorithms that historically have not served them well in areas such as predicting recidivism, which is the rearrest, reconviction, and cycling back of an offender into the criminal justice system after being released on parole or probation.[10] AI-based recidivism prediction algorithms are often used by judges in the process of deciding who can be set free, or how much should a bail bond be. In essence, these algorithms use machine learning to score the likelihood of recidivism and have been accused of being biased against Black individuals.[11] Alternatively, their lack of adoption could be due to a lack of awareness or understanding of the possible benefits these algorithms can provide. *But not adopting AI results in a significant economic drawback: sacrificing an 8.6 percent increase in revenue.* This downside was higher for Black hosts who gained significantly more demand for their properties when they used smart pricing.

While on the surface, listing the Jaipur flat on Airbnb may primarily have looked like an economic decision, in reality it was also a way for them to "take care" of Ravi's aging retired parents by giving them some purpose. Acting as the local hosts for the flat, his mom and dad, Usha and Jawahar respectively, got to meet travelers from all over the world. These human connections became very meaningful for them both, but especially his father, a retired pharmacology professor who'd been a postdoc researcher at the

National Institutes of Health in Bethesda, Maryland, and missed the social interactions of his previous career.

For elderly Airbnb hosts such as Ravi's parents, these human connections become very meaningful. As they age and travel becomes more difficult, seniors make fewer trips. Acting as Airbnb hosts helps counter this. If an elderly couple or individual operates a rental property, it brings the world to their doorstep in the form of travelers who converse with them, ask for their recommendations for restaurants and things to do in the city. These guests not only interact with older hosts socially but also help them feel valued and give them a sense of purpose.

There is more to the Bapnas' Airbnb hosting experience. The following is the kind of message that Ravi's parents occasionally receive through Airbnb. This specific message was from a guest seeking to book their place for a one-night mid-week stay (the message shared is verbatim, with actual typos but with scrambled numbers, from a guest we will call Person A):

> Person A: "Sir please share your contact details I want to book it or you can call me on nine one one six zero one two seven six."

Consider next the resulting sequence of exchanges between Airbnb, Ravi, and Person A (as recipient).

Person A had zero reviews and was violating a key rule for guests and hosts at Airbnb by engaging in communication outside the platform, and, furthermore, using trickery to share his phone number (writing them as words instead of using numerals). While Ravi and Sofia are experienced superhosts who see through situations like this and flag them (again, just as Gmail encourages you to tag junk mail as spam and use that as future training data to refine its spam-detection algorithm), it is possible that newer hosts are susceptible to great risk from guests like Person A. This is where a cutting-edge AI—in the form of *intent analytics*—once again comes to the rescue. Here, the guest is being opportunistic at the very least and potentially fraudulent (for example, by staying but not paying or not being subject to reviews) in violating the platform's rules. So, what is intent analytics and how does it protect hosts in such situations?

Airbnb's AI team combines ideas of unsupervised learning, supervised learning, and natural language processing—three AI and ML concepts with which you are now familiar—to detect the *intent* of any conversation happening in their messaging system. The first step in building this AI is to understand the various *topics* that people might discuss. There are always conversations about the coordination of arrivals, preferences around extra services such as housekeeping frequency, prebooking inquiries about various facilities such as the pool or restaurants, and, fortunately rarely, the odd occasion where the intent is to go outside the platform and negotiate a "deal." This is where an unsupervised AI model that works with natural language such as *latent Dirichlet allocation (LDA)*[12] comes in handy. Given a large corpus of text—say all Airbnb conversations over the last five years—it can detect the *topics* being talked about in every conversation. A subset of these same conversations is then also read and *labeled* by a team of product specialists as ones with a particular intent, such as an "illegal offline communication" or "customer support" or "payment problem." Yes, we can have multiple categorical levels (more than two)—10, 20, or even 200 (this is similar to detecting objects such as tree, desk, river, book, etc. in a photograph)—as classes in our outcome variable. With topics as inputs and labels as outcomes, now we can deploy supervised machine learning to take any new conversation, detect the likelihood of illegal communication in it, and raise a flag,[13] as was done in the bold-underlined text in the conversation shown in figure 3.1.

```
Airbnb:     "Looks like there is an attempt by the guest to
communicate outside the platform. Is this correct? YES or NO"

Ravi: YES

Airbnb:     "Thank you for flagging the user for violating the
platform's rules."

Ravi presses "Inquiry declined"

Ravi: "You need 3-5 positive reviews to book our place as
outlined in our listing and you cannot contact us outside of
Airbnb."
```

Figure 3.1

Airbnb conversation with malicious intent detection (in bold and underlined) and subsequent response (in bold).

This exactly what happened with the previous message where Ravi and Sofia received an immediate notification from Airbnb asking, "Did the guest try to communicate offline?" Ravi declined the booking inquiry and reminded the guest of Airbnb's rules. The same intent analytics method is used to reduce other frictions in the marketplace, such as predicting customer support issues by leveraging message intent history and providing immediate responses to guests' and hosts' needs.

Hopefully, you're starting to notice a common thread across these stories. We saw Jasmina protected by her bank against fraudulent transactions using anomaly detection and the role of deep learning in detecting lewd images or unsavory messages on OkCupid, Alyssa's doctor relying on heart disease predictions from supervised machine learning and recommendations from reinforced learning when she bought her new camera, and now this new situation with Ravi and Sofia concerning opportunistic user behavior on Airbnb.

The AI technologies of today can keep us safe. Period.

Stranger danger aside, we have a feeling that Airbnb's founders would enjoy the story about Ravi's parents becoming enthusiastic hosts to visiting travelers. Brian Chesky and Joe Gebbia have written about their vision for the company and spoken of it in interviews. Their optimism shines through in their words and suggests that they are on mission to do more than simply make money off room rentals.

In July 2022, Joe Gebbia announced that he'd be reducing his involvement in the company's operations, citing the imminent arrival of his first child. He shared his announcement in a letter that was published on the Airbnb news site. In the letter, he recalls another birth. In 2007 he and his roommates started the company that would become Airbnb after their landlord raised the rent on their apartment. He says: "Incredibly, after a billion-plus guest arrivals, the data proves that the Golden Rule is actually human nature, which is perhaps why some version of it can be found in almost every culture. In other words, people are good. Sometimes with all the stories of suffering in the world, we need to be reminded of that."[14]

On Twitter, Brian Chesky shared a letter he wrote to the Airbnb team about his partnership with Gebbia. It culminates with this paragraph:

> One day I asked Joe, "What would success for Airbnb be?" You know what he said? It wasn't to become the biggest company in the world. It was to expand the definition of the word "family." He said that if we were successful, you would open the dictionary one day and in the dictionary the word "family" wouldn't be limited to your parents, siblings, or children. It would also include all the people you took into your home—the people you cared for, travelers that you were entrusted with.[15]

IS IT REALLY SO BAD?

Some might dismiss the words of Airbnb's founders as platitudes. Maybe so, but it's harder to dismiss the reach of Airbnb. As Gebbia put it in his letter, "Thanks also to the four million of you who've chosen to host, defying convention to form the largest network of hospitality on the planet. You've welcomed the world into your guest bedrooms, yurts, villas, caves, barns, mansions, tugboats, Airstreams, and that one giant potato house in Idaho."[16]

We suspect there might be another reason why Gebbia and Chesky highlight the positives their technology and business have generated. Digital platforms, apps, and other AI-enabled technologies have become central to twenty-first century life throughout much of the world. Along the way, these innovations have been scrutinized in the media and elsewhere. More often, the media, popular culture, and politicians focus on negative experiences or bad outcomes: the app that facilitated bullying among teens, the platform that allowed trolls to go unchecked, the AI devices that felt a bit invasive and creeped people out, the bots that influenced major elections. When it comes to Airbnb specifically, there has been media coverage of the supposed negative impacts on nonhosting residents. Essentially, this narrative claims that Airbnb rentals cause headaches for neighbors who live near an Airbnb listing, presumably because renters bring noise, parties, parking problems, and general disrespect to the places they stay. For example, in 2016 the *New York Times* ran an article with the provocative headline "Airbnb Pits Neighbor Against Neighbor in Tourist-Friendly New Orleans,"[17] which states, "The technology design that has disrupted the hospitality industry has also disrupted civic life and public policy making." And that disruption is spurring debate about short-term rentals in cities like New Orleans, as well as Portland, Oregon, and Austin, Texas. The article does mention that there is a mix of opinions among New Orleans residents, including some who cite the influx of rental dollars as a means to restore buildings and communities damaged by hurricanes, but the tone of the article tilts to the negative, and it paints Airbnb and other similar companies as remote entities that do not want to get involved with the local communities where they operate. The implication is that Airbnb and its hosts are operating unregulated businesses without oversight or proper licenses and in violation of local ordinances.

However, empirical research from 2021 finds that "Airbnb's perceived negative impacts are not as negative as portrayed in media discourse."[18] The study surveyed 415 residents who live in areas with Airbnb activity. The survey was designed to collect information on the residents' attitudes and perceptions of tourism in general and the negative and positive impact of Airbnb in particular. The idea that short-term renters could be disrupting neighbors and neighborhoods sounds plausible on the surface, but the researchers' findings showed "that Airbnb is perceptibly inconsequential enough to a large number of residents, particularly the average resident captured in the present study, to enhance or deteriorate their [quality of life]." Essentially, the surveyed residents did not perceive Airbnb rentals as having much effect on their lives at all.

Of course, history shows us that technology of any kind can have positive, negative, and unintended effects, especially when it's new and people are adapting to the changes it brings. The introduction of new technologies is often accompanied by hand-wringing worry among some parts of the population. Several examples of this kind of "panic" come to mind easily, including the introduction of the automobile, video games, headphones and even the pocket calculator. All these technologies were once vilified as harbingers of certain doom for humanity.

Like those earlier technologies, digital platforms, apps, and artificial intelligence are taking their turns causing panic. They aren't perfect or 100 percent positive in impact 100 percent of the time, but neither are they all bad. Our perspective is that digital platforms, apps, and AI often bring out the best in people rather than the worst. These positive stories don't receive as much attention in the popular press, perhaps because people are severely prone to negativity bias or because stories of technological doom and gloom garner bigger ratings and more clicks. Sadly, fear and negativity sell big.

Becoming Airbnb hosts brought new friends and human connections to Ravi's parents at a time in their lives when they might otherwise have drifted toward isolation and loneliness. They invited renters to share meals with them and shared secrets for shopping the old bazaars of Jaipur. In our view, a technology that fosters such positive human connections cannot be written off as inherently detrimental. After all, even the humble hammer can be used for both building up or tearing down.

More interestingly, the adoption and use of AI by other digital platforms in the sharing economy is becoming rampant. DeepETA is Uber's most advanced algorithm for estimating arrival times using deep learning. Used both for Uber and Uber Eats, DeepETA organizes everything in the background so that "riders, drivers and food are fluently going from point a to point b in time as efficiently as possible."[19] Uber uses AI for fraud detection, risk assessment, safety processes, marketing spending and allocation, matching drivers and riders, route optimization, driver onboarding, and in several other domains.[20] On the Lyft app, interactive help is a chat-like experience, which is powered by AI and machine learning-powered bots. AI is used to predict what questions a customer might have and then personalizes the experience for the passenger while resolving the issue at hand.[21]

BEYOND RATINGS, TOWARD TRUST

Obviously, building the Airbnb platform took significant talent, effort, and resources. Timing also played a role, as it often does when it comes to ideas and tech. Imagine trying to build Airbnb using the tech of the 1990s or even during the dot-com bubble. The company might have never gotten off the ground. Luckily, the founders came together at a time when Web 2.0, social media, and smartphones were quickly remaking how people communicate, connect, shop, and yes, travel.

Certainly, Airbnb faced countless barriers before the company became a household name and the de facto synonym for private rentals. Some of these barriers were no doubt engineering problems and technical in nature. Others were of the messy human variety, including the human need for trust. Opening one's home to strangers—or being willing to stay overnight in the home of someone you don't know, perhaps in a place you've never been before—requires enormous trust. Both parties need to have enough trust in the other to feel comfortable and secure in the situation. More than anything else, an Airbnb booking hinges primarily on human trust.

Airbnb didn't invent room rentals. People have been sharing accommodations for centuries, with acquaintances and strangers, in all kinds of ways. Staying at a boarding house, youth hostel, or a bed and breakfast all touch on the same need for trust between renter and proprietor. So too would the act of posting a classified ad for a roommate and interviewing the strangers who responded. But Airbnb scaled rentals beyond what anyone had ever done before and addressed the trust question in ways that normalized the idea for millions of people. What happened to make so many people from different cultures around the world willing to trust strangers so readily?

Airbnb, Uber, Lyft, TaskRabbit, and other companies built on connecting individuals for services and transactions must all deal with the question of trust among users. Ratings, reviews, and a recognizable brand name certainly play a part in establishing trust, but those mechanisms alone may not have been sufficient for a company like Airbnb.

Ratings and reviews have long been a part of consumer decisions. In the twentieth century, endorsements from the Good Housekeeping Seal of Approval and Consumer Reports offered buyers the assurance that experts had vetted, tested, and recommended a product over the competition. Countless celebrities have acted as spokespeople for all kinds of companies for decades (if not centuries: "Aristotle swears by this toga maker!"). These practices continue today, but what is relatively new are the ubiquitous online reviews from individuals willing to share their thoughts about purchases. Personal computers, databases, online payments, and social media have made this possible, shifting more of the ratings and reviews game to individuals and not just tastemakers and gatekeepers. Thanks to YouTube and other twenty-first-century platforms, product reviews have become a cottage industry, one in which a kid who shares opinions about toys can make millions of dollars.[22]

But buying a vacuum cleaner or paperback based on ratings and reviews is much different from renting a room in someone's house or getting into the backseat of a car with a stranger at the wheel. This is no doubt why companies like Airbnb went beyond the basic idea of using ratings and reviews to signal quality and reliability to newcomers. Airbnb's efforts to build trust among users includes a multidimensional approach that allows users to self-select into what aspects of trust matter most to them. Users use their own prior interactions as a proxy for future trust. Each time we buy from an eBay seller and the product shows up at our door, our trust in the seller increases. If you are on TaskRabbit you might select a *tasker* who has completed a task for you before. It can also come from the experience of others, typically in the form of a review. Further algorithmic flagging of problematic buyers and sellers provides capabilities that are difficult to match in the physical world. Airbnb uses advanced text analytics to actively moderate prebooking chat conversations, putting up guardrails that protect both buyers and sellers. The evolution of trust as a core component of business, commerce, and capitalism is a fascinating subject, one that Arun Sundararajan examines in his book *The Sharing Economy: The End of Employment and the Rise of Crowd-Based Capitalism.*

Whereas people once relied on reputations, handshakes, and signatures as indicators that a potential business partner could be trusted, today's signals are digital in nature. Online information now plays a key role in helping individuals understand which people and companies to trust. Profiles on social media networks like Facebook, YouTube, TikTok, Pinterest, Twitter (X), Snapchat, and LinkedIn, reviews on sites like Yelp and Tripadvisor, third-party certifications, endorsements, and other digital cues can all be useful in signaling legitimacy. Such digitized social capital now plays a key role in greasing the wheels in fostering human connections in the exchange of goods and services. Modern AI-powered platforms actively try to maximize the feeling of interpersonal connection. We mistrust people we don't know. This is done using dynamic nudges for people to interact, by getting the size of the text box to be just right. Too small, and it will induce a "What's up?" feeling, and too long, it won't be read or will induce too much information, negatively impacting the likelihood of a pleasant interpersonal experience.[23]

This evolution has happened fairly quickly. In under a decade, many of us went from not having a Facebook profile to eagerly accepting rides from strangers because a smartphone app made us feel safe enough to do so. A turning point in this evolution came from one of the earlier Internet giants: eBay.

Over the last twenty years or so, platforms such as eBay and Amazon Marketplace have facilitated millions of stranger-to-stranger transactions every month. Buying a collectible, antique, or other item on eBay may not require the same level of trust as a room rental, but there is financial risk involved. Winning an eBay auction requires buyers to submit payment to sellers before receiving the purchased item. When it does arrive, the item might be inauthentic or of lower quality than described in the listing. Without trust, it's unlikely that millions of people would participate in eBay auctions, especially when they could easily shop at a local retailer with a traditionally established reputation. eBay, however, operates "one of the earliest and best-known Internet reputation systems" on its platform, and, according to a paper by Paul Resnick and Richard Zeckhauser, "trust has emerged due to the feedback or reputation system employed by eBay and other auction sites."[24]

In their work, Resnick and Zeckhauser examine how eBay's reputation system operates. They note that computer-enabled tools made the collection and distribution of evaluations easy and cheap for eBay participants. This streamlined availability of information facilitates trust building in a different way compared to traditional means, but it has proven effective online. Although Resnick and Zeckhauser's analysis reveals flaws in such reputation systems, they conclude that those flaws may not matter in the end. Essentially, the system works because participants *perceive it* to signal trust.

Airbnb's approach builds on the model eBay pioneered. Airbnb incorporated mechanisms such as the ability to rate a property and write a review, along with other important digital signals. As Sundararajan explains in his book, a trail of digital signals, including things like an active Facebook profile, give people a way to gauge the authenticity of another person. An important component of Airbnb's design is that it requires users on both sides of the transaction to verify their identity with a Facebook profile. This affords users the ability to "check out" a potential renter or host online before entering into a transaction, especially one as intimate as sharing your home with a stranger.

WHEN TRANSPARENCY LEADS TO DISCRIMINATION

Airbnb's founders may project an outlook and vision for the company that is relentlessly optimistic, but human nature is not always so. The information transparency that engenders trust among users may also prompt discrimination on the part of individual actors using the platform. This has been shown in several contexts, both within and outside of the sharing economy.

For example, field experiments have examined how ridesharing apps like Uber, Lyft, and others may facilitate discrimination, even when platforms implement design changes intended to prevent such behavior. In a 2017 paper on the topic, Jorge Mejia and Chris Parker explain that "while the algorithmic matching of these two-sided platforms may be efficient based on specific objective criteria, the behavior from the humans executing the service delivery after matching may not."[25]

In other words, individual users who sign up to participate on the supply side of such platforms (e.g., driving for a rideshare company or listing a room to be rented) have been shown to behave in discriminatory ways. In one experiment, Mejia and Parker show that rideshare drivers cancel pickups of customers from some groups more often than others. Notably, the study shows that racial discrimination persists in rideshare platforms, even after the companies operating those platforms try to mitigate the effects by withholding some information from drivers until later in the transaction. The paper also documents that discrimination occurs when riders are perceived to signal support for lesbian, gay, bisexual, and transgender (LGBT) rights and causes. (The experiment did not uncover a discriminatory effect based on gender alone.)

Uber has come under fire in the mainstream media for discrimination. A former Uber driver in San Diego sued the company for racial discrimination regarding how it uses passengers' reviews to evaluate drivers.[26] Apparently, some riders would cancel requests after the driver had already accepted the ride and the rider was able to view his picture. Another academic study conducted in Seattle and Boston looked at the results of nearly 1,500 rides hailed via Lyft, Uber, and Flywheel on controlled routes and showed that African Americans wait longer for rides in Seattle.[27] Yet another academic study found that Black people received more negative reviews than white workers in online labor marketplaces such as TaskRabbit and Fiverr and that female workers were less likely to receive any feedback at all.[28]

Similarly, other researchers have found discrimination present among Airbnb hosts. A large-scale experiment on the platform in 2017 tested roughly 6,400 Airbnb listings in five cities. The researchers used invented user accounts to inquire about booking availability of Airbnb listings. Some accounts displayed a "distinctly African American" name and others a "distinctly white" name; otherwise, the accounts were identical.[29] The study showed that the "distinctly African American" accounts received fewer positive responses from hosts as compared to identical accounts with "distinctly white" names. The authors quantify the difference at 16 percent and note, "The penalty is consistent with the racial gap found in contexts ranging from labor markets to online lending to classified ads to taxicabs." Namely, some renters are less welcomed by hosts than others, despite Airbnb's sunny outlook.

Discrimination can run the other direction as well. In another recent study, researchers examined the listing prices for 100,000 Airbnb hosts in fourteen countries. According to their analysis, "Compared to White hosts, Black hosts charged 7.39 percent lower prices and Asian hosts charged 5.94 percent lower prices for similar apartments."[30] Although the exact numbers varied, they found the differences to be widespread in the cities they studied, with white hosts consistently listing properties at higher rates. They note, "These findings support the hypothesis that, all else being equal, consumers prefer to stay with White hosts, which allows them to charge higher prices."

The findings from experiments like these and others represent a threat to platforms like Airbnb. Negative press coverage and social word of mouth among users experiencing discrimination have the potential to damage the company's reputation and worth, at the very least. One can imagine the more severe consequences that could occur for minority users should this behavior continue unchecked on the platform.

In response to the evidence, these platforms have taken action. Uber, in its attempt to prevent discrimination based on someone's name, has allowed riders to update their names in the app. Uber and Lyft have responded to stories of discrimination against riders by firing drivers and altering the app.[31] In recent years, Airbnb has assembled teams to propose features and design changes to the reservation interface, in addition to directly assisting users who report discrimination.[32] Some of the design changes under consideration revolve around information on renters and hosts as well as what is shared and when. For instance, in 2022 the company started testing the effects of using only a user's first initial instead of full name until the booking is confirmed.[33] The aim of an experiment like this is to see the effects on discrimination rates in an attempt to isolate the mechanisms that can promote prosocial behaviors like trust or deter negative behaviors like discrimination.

As researchers, we know the power of such experiments, commonly referred to as *A/B testing*. You've likely encountered the concept in any number of contexts and fields that rely on the scientific method. The design of such experiments is to assemble two identical groups, treat only one of those groups with some kind of intervention, and then study what happens. Doing so allows researchers to determine whether the intervention is actually responsible for any differences in outcomes. In other words, one thing can be shown to *cause* a particular outcome.

A/B testing is not new, but the use of A/B testing in service and product design is relatively recent compared to fields such as science and medicine. This is likely because running large experiments with real customers can be costly and logistically unwieldy. The advent of data-driven digital businesses is changing that. Companies such as Microsoft, Meta, TikTok, Twitter, Uber, Google, Amazon, and yes, Airbnb, are operating extensive testing programs that examine minute changes to product features. "We use controlled experiments to learn and make decisions at every step of product development, from design to algorithms. They are equally important in shaping the user experience," writes Jan Overgoor, an engineer at Airbnb.[34]

As we stated in chapter 1, rigorous testing is essential to succeeding in an AI-powered digital world. It is not only important, but also indispensable when aiming to identify causation. But we don't believe in turning over all decisions to machines and algorithms. Human perception and judgment must also play a role, as described in our House of AI framework.

WE ARE MORE CONNECTED AND MORE MOBILE THAN EVER

Digital platforms such as Airbnb, Uber, Lyft, eBay, Etsy, TaskRabbit, and many others have made it possible for individuals to connect directly with one another to obtain a wide range of products and services. The effects are playing out around us all the time, in an exponential pileup of transactions, exchanges, and interactions.

At a microlevel, some of these interactions might seem mundane. The sale of a gadget on eBay might seem inconsequential at this point in history, but the impact may be profound it if helps a seller in a developing country connect to the broader world. So too, Etsy is helping artists not only gain access to a wider market but also connect with art lovers who want to bring more beauty into their lives. Ridesharing companies are providing people with a more flexible form of transportation. And Airbnb is making it possible for people to meet and even bond with strangers in another part of the world. Could any of these things happen without the technology, data, and capabilities of a digital world?

And now, in the wake of the COVID-19 pandemic, these trends seem likely to continue and even multiply. Months of lockdowns around the world in 2020 and 2021 proved to companies and workers that physical presence in an office building is not always necessary for many kinds of work. Observers and researchers (including Airbnb itself) are already seeing more people take advantage of this increase in flexibility.

"For those fortunate enough to be able to work remotely, working from anywhere has become a viable lifestyle," states an Airbnb report from 2021.[35] The same analysis shows that people feel able to consider a wider array of options, both in terms of dates and destinations. They're traveling to a larger variety of places and staying longer. The report notes, "The percentage of long-term stays (at least 28 nights) on Airbnb almost doubled from 14 percent of nights booked in 2019 to 24 percent of nights booked in Q1 2021."

If more people start traveling this way, then who will end up meeting whom? What connections will be formed as more people experience cultures other than their own and for longer periods of time? What opportunities will be uncovered as more people explore more places on the planet we share?

We're about to find out.

TAKEAWAYS

- AI is fostering human connections between strangers in a variety of ways. Academic research has shown a significant downside from not adopting AI algorithms such as smart pricing on Airbnb. Nextdoor uses AI and ML to predict and detect, in real-time, when a comments thread may become heated *before* a neighbor contributes. When the AI model predicts a high risk of a thread getting nasty, the app intervenes with a reminder and some tips on how to write with a more empathetic tone.
- The AI technologies of today are keeping us safe in many ways. Just as banks can keep us safe from fraudulent transactions using anomaly detection, deep learning keeps us safe by detecting lewd images or unsavory messages on dating apps; doctors can rely on heart disease predictions from supervised machine learning and recommendations from reinforced learning; and AI algorithms can perform intent analytics using topic modeling to prevent opportunistic and fraudulent behavior in platforms that connect people.
- AI algorithms are alleviating and often preventing discrimination in platforms. The information transparency that engenders trust among users may also prompt discrimination on the part of individual actors using the platform. This has been shown in several contexts including Airbnb and Uber. As we stated in chapter 1, rigorous causal analytics (A/B testing) is essential to succeeding in an AI-powered digital world. In response to the evidence of discrimination, companies such as Microsoft, Google, Meta, Uber, Amazon, and Airbnb are operating extensive testing programs leveraging the same causal analytics pillar to examine minute changes to product features and improve the overall user experience.

4 AI, MHEALTH, AND THE QUANTIFIED SELF

You may have heard the phrase "data is the new oil." The book's authors often preach and practice a corollary of this: "great data trumps great models" (of the AI and ML kind). What do these phrases mean? "Data is the new oil" means that data is the truly valued commodity in our new digital society and is worth a great deal (although unlike oil, we won't run out of data). "Great data trumps great models" means that we as researchers can build all the hypothetical models we want, but nothing beats having truly unique and insightful data to analyze trends, causally explain insights, and predict future results.

So far, we have showcased the power of various forms of AI models that help us refine data to produce novel insights. What does the AI-powered digital ecosystem have to offer with respect to the vibrancy of the ultimate and still primarily not understood machine called the human body? Consider the story of Hazel, a (fictional) thirty-year-old English teacher at Langdon High School in Cavalier County, North Dakota. If you have not heard of Cavalier County, a largely agricultural community bordering Canada with a population of 3,704 residents, you are not alone. While the local chamber of commerce touts the county's family values and proximity to nature, it fails to mention that it is one of the approximately 35 percent (a shockingly large fraction)of counties in the United States categorized as maternity deserts. A maternity desert is a county with no obstetric doctors and no birth centers that offer obstetric care.[1]

Hazel is eagerly awaiting the birth of her first child, and her parents are over the moon at the prospect of becoming grandparents. As a baby shower gift, they gave Hazel a top-of-the-line WHOOP fitness band and a year of its monthly membership to use its data-rich app. The WHOOP has multiple sensors that collect data 100 times per second and produce close to 100 megabytes of data per day, per user. This is significantly more than other fitness trackers such as the Apple Watch or Fitbit, and the WHOOP band is the favorite of superstar athletes such as Clippers' DeAndre Jordan.[2]

WHOOP's value proposition is that it helps its users calibrate their strain, recovery, and sleep levels, which is something other trackers can't do. Too much strain results in a poor night's sleep, and a poor night's sleep indicates a poor recovery, which in turn might be motivation for users to clean up their sleep hygiene habits. A key parameter at the heart of these three zones of strain, recovery, and sleep is heart rate variability (HRV), and research shows that HRV is notoriously hard for wrist-based devices to pin down, especially when people are exercising.[3] The appeal to Hazel's parents is that the WHOOP also has pregnancy coach functionality that can alert Hazel of any abnormalities that might put her or her baby at risk as her pregnancy progresses.

Hazel is in her last trimester, and last week (week twenty-nine out of forty) was unusually uncomfortable. She hasn't seen anyone about her discomfort because it's an almost three-hour drive to Fargo to see the doctor she really likes and, at best, telehealth services only connect her with nurse practitioners, who have not been much help in the last few weeks. She pushes through another week of class, but it's a strain since her students are needier and her workload is heavier with mid-term exams and projects as the semester draws to a close. Her WHOOP shows that her strain levels are not going down and her recovery levels are not going up, which isn't great, but what freaks her out is an alert from the app that detects an abnormal uptick in her HRV. Her doctor says that for a full-term pregnancy, her HRV is supposed to steadily decrease until week thirty-three and then steadily increase[4] (see figure 4.1), and if this doesn't happen or happens too early, she might be at risk for a preterm delivery. WHOOP's pregnancy coach feature detects Hazel's abnormality and alerts her that this HRV uptick is happening two to three weeks earlier than it should for a full-term pregnancy.[5]

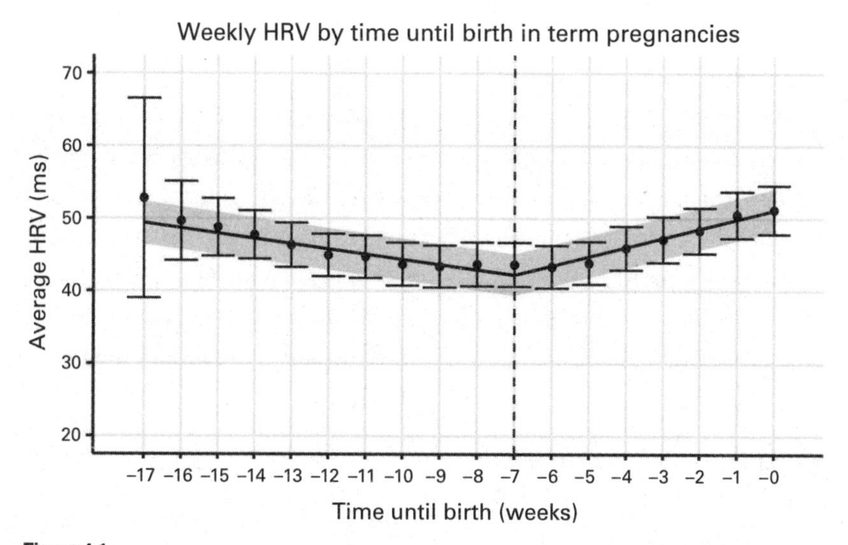

Figure 4.1

A new data source reveals a new marker to detect premature delivery. HRV should steadily decrease until about seven weeks from delivery, and then it should steadily increase. If this happens earlier than week thirty-two or thirty-three it could be indicative of a preterm delivery, which normally requires special medical attention. Monitoring patients' HRV allows doctors to then act accordingly to care for potential preterm births. *Source:* Summer R. Jasinski, Shon Rowan, David M Presby, Elizabeth Claydon, and Emily R Capodilupo, "Wearable-Derived Maternal Heart Rate Variability as a Novel Digital Biomarker of Preterm Birth," *MedRxiv (Cold Spring Harbor Laboratory)*, November 5, 2022, https://doi.org/10.1101/2022.11.04.22281959.

Rightly concerned based on her HRV numbers, Hazel sought and was able to get proper care in Fargo for her preterm delivery (and she *did* deliver early). In the absence of an AI-powered digital ecosystem, how many other women are deprived of this benefit in the vast maternity deserts of the United States where no care is nearby? (Remember, it's 35 percent of the country.) Just imagine many preterm deliveries could be prevented with wider access to this technology.

This is a classic case of data being the new oil and great data trumps great models in one setting. In this case, WHOOP's richer sensor data (100 measurements per second) led to better accuracy in continuous monitoring of Hazel's HRV (supervised learning also played a role in improving this accuracy[6]). Thus, a simple trend chart (no fancy machine learning model was needed to create figure 4.1) that Hazel's doctor saw and shared ended up potentially saving two lives!

In chapters 2 and 3, we explored how today's AI-powered digital ecosystems are converging with machine learning and artificial intelligence to impact our emotional lives. Finding love and making human connections (or not) certainly affect humans in tangible ways that go beyond emotional states. Next, we examine how AI tools affect our physical health, where the same forces are converging to improve the health of individuals and populations. In this chapter, we explore some of the ways AI is reshaping our health for the better.

ALEX, THE TRIATHLETE

Remember Alex, Jasmina's hiking date from chapter 1? Alex is an athletic guy. He grew up participating in team sports through community recreation leagues. In high school, he joined the swim team. That commitment required regular workouts in the weight room as well as the pool. In those days, Alex recorded his weightlifting progress and race times in a pocket notebook.

As it turns out, this book's authors are both fitness-focused individuals. We love to buy various kinds of fitness gadgets and athletic gear on Instagram, which does a remarkably precise job of targeting us with the right ads for the right products. Furthermore, Anindya trains for ultra-endurance activities through most of the year because of his passion for high-altitude mountaineering. We'll offer more on Anindya's mountaineering background in a bit; for now, let's return to Alex.

In college, Alex joined a running club on campus. He enjoyed the chance to get some exercise, meet new people, and blow off steam. During runs along routes the club had mapped out, Alex used a digital watch to time himself (this was in the late 2000s before modern fitness trackers became commonplace). Back in his dorm room, he did some simple math to calculate his pace.

Over time, Alex moved from the more manual fitness tracker to one tied to a smartphone app that recorded and stored all the data he needed automatically. By signing up for the app with his social media profile, Alex was able to connect with his friends, training partners, and other triathletes in a new way. The app allowed users to share workout stats and routes with one another. These posts appeared in Alex's social media feed, inspiring him to push himself further during his next workout. When he posted about his own achievements, others left congratulations and notes of encouragement in the comments. These exchanges helped Alex stay motivated to keep up with his training, and he felt supported even when he worked out alone.

When the Apple Watch appeared on the market in 2015, Alex was an early adopter. Wearing a smartwatch made it even easier to track his workouts and performance. Not only did the watch collect data on his physical activity, but it also tracked his sleep patterns, heart rate, calorie burn, and blood pressure. The watch and its accompanying apps suggested customized workouts based on Alex's data.

Recently Alex upgraded to a newer-model smartwatch to take full advantage of the device's biometric capabilities, new waterproof design, and other features. He also bought an entry-level model as a gift for his father. Alex liked the idea that his father—eighty years old but still living on his own in a single-family home—would get reminders to take his medication. And if he took a bad fall, Alex's father could use the watch to call for help. In fact, if his watch detected a fall and his father didn't respond, it would call for help *for* him.

Alex is one of the millions of people using AI-powered digital ecosystem to monitor, track, and improve his health and personal wellness. And whereas few of us will train for triathlons like Alex, many have found use for what is called the *quantified self*—the concept that collecting data on one's eating, sleeping, and physical movement can provide insight into the state of the body and lead to greater health, fitness, and overall well-being.

WELCOME TO THE MHEALTH ERA

As we've described, Alex uses various and evolving fitness-related tech to collect biometric data on his physical activities. The point is to learn from the data and make meaningful changes. In Alex's case, when he has enough data, he can study it and, based on what he sees, make changes to his training regimen.

This is not a new idea, but it is worth noting how far the potential for changing personal health behaviors based on AI data has come in a relatively short period of time. We both have first-hand experience with this. Ravi, as we learned in chapter 1, has a family history of heart disease, so he uses an Apple Watch to monitor his ECG so he can keep his LDL cholesterol at a healthy level. Anindya, as mentioned, is a high-altitude mountaineer. He has climbed in the Himalayas, Andes, Alps, Cascades, and Rockies, as well as well-known single peaks such as Kilimanjaro. He has been hiking, trekking, and climbing for almost twenty-five years, which means he has trained continually through those years. The amount of oxygen in the bloodstream decreases significantly at higher elevations, causing a variety of negative symptoms. Because of the lack of oxygen at altitude, it is more difficult to digest food, there's a loss of appetite, we become dehydrated quicker with the lower pressure, we breathe more rapidly for days until we acclimate, and there's some loss of cognitive ability. All of this makes climbing a very demanding sport, and high-altitude mountaineering in particular requires multiple different kinds of conditioning such as cardio, interval, strength, flexibility, endurance, and balance. This makes the data science behind it even more critical for supreme performance at high altitudes where the lack of oxygen can be devastating.

Twenty-five years ago, climbers like Anindya would just wear a Casio sports watch that showed them the current altitude and barometric pressure. About ten years ago, smartwatches from the likes of Apple and Samsung began to give us data on distance walked, elevation gained, heart rate, and calories spent. As of 2023, the latest Apple Watch gives us blood-oxygen saturation levels and body temperature readings. Climbers now regularly use wearable mHealth (mobile health—more on this later) technology to track their statistics while ascending the world's most perilous mountains. The value of wearables for climbers is immense because they offer hands-free technology that can be used while they climb. Smart wristbands monitor movement, muscle exertion, and speed, and the resulting data can be used to customize training regimes and highlight areas for improvement. Telemetric clothing uses AI to monitor body metrics, track how well athletes perform, understand their body physiology, and then recommend how they can adjust to perform better.[7]

For example, during Anindya's climbs and hikes in the Andes in Bolivia, Chile, and Ecuador, he and his fellow climbers not only relied on various trackers with smartphone apps to track various physiological performance metrics, but also on Bluetooth-enabled clothing embedded with sensors that continuously track heart rate, oxygen saturation, skin temperature, altitude, and location. These technologies enable not only the participants to self-monitor, but also the support teams at base camp to monitor each climber's health in real-time, watching for signs of edema, hypothermia, and cardiac issues in order to predict a health concern in advance and intervene before an emergency occurs. The AI technology underlying these data analyses relies on the second pillar, predictive analytics, of our House of AI framework from chapter 1.

Within a couple of decades, Anindya went from recording a limited amount of information using paper and pencil to wearing a device (and even clothing!) that automatically and continually collects a growing range of data points with much greater precision and records them in an app that can be accessed from a smartphone, tablet, or desktop—usually all three, depending on the user's preference. Instead of estimating the number of floors he climbed or the pace and distance he ran, Anindya can gather real, accurate data, day after day, workout after workout. Moreover, he doesn't have to think about it very much: technological advances in cloud computing, network connectivity, GPS tracking, sensors, data storage, and software apps have converged in a sleek, elegant device that can be worn on the wrist. (We hear it also tells the time!)

As we can see in the example of high-altitude climbers, such devices hold the potential to be much more than fun gadgets. Smartwatches, fitness trackers, and other wearables have become part of a disruptive, digital transformation of the healthcare industry. There are even app-connected smart bottles designed to track water intake and remind users to hydrate.[8] As far back as 2012, BodyMedia (which was acquired by JawBone in 2013) announced that its FIT Armbands were being used on the International Space Station as part of a study to determine astronauts' energy needs during long space flights.[9] This trend, as facilitated by AI-powered smart mobile computing technologies, is often referred to as *mHealth*—for mobile health—and it is helping push the healthcare system away from reactive care and toward proactive prevention and intervention.

The term *mHealth* broadly refers to the various ways mobile computing, medical sensors, and communications technologies are being combined to deliver healthcare in new ways. This includes medical software that runs on smartphones and tablets, sensors that track vital signs and health activities, and cloud-based computing systems that collect data of interest to medical professionals. The global mHealth market size was valued at $50.7 billion in 2021 and is expected to expand at a compound annual growth rate (CAGR) of 11.0 percent from 2022 to 2030.[10] Increasing awareness of the utility of mHealth apps for remote patient monitoring, patient tracking, medication management, disease management, fitness and wellness, women's health, and personal health record management is anticipated to drive growth over a number of years.

Examples of how AI and mHealth are changing the healthcare landscape—and people's lives—are plentiful and range from the basic to the boundary-pushing. Not surprisingly, given the advances in image recognition using deep learning, an advanced prediction modeling approach, some of the breakthrough practical applications of AI are in radiology. In fact, Geoff Hinton, one of the pioneers of deep learning, told the *New Yorker* in 2017, "I think that if you work as a radiologist . . . you're already over the edge of the cliff, but you haven't looked down."[11] While Professor Hinton was prescient about the accuracy of AI to detect, for example, breast cancer, surpassing humans, we see AI as being an important augment to the capacity of the trained radiologist, rather than a replacement. This view was validated by a 2023 headline story in getting *New York Times* that took us to a small breast cancer clinic in Budapest called MaMMa Klinika.[12] Here, AI is used to review and validate breast cancer diagnoses made by two staff radiologists. This AI tool often agrees with the doctors, but in several cases, it also flagged areas of mammograms that the doctors missed. According to the story, across five MaMMa Klinika sites in Hungary, twenty-two cases have been documented since 2021 in which AI identified a cancer missed by radiologists, with about forty more cases under review. Think about how many lives this technology is saving on an ongoing basis.

Other studies have examined the use of mHealth in chronic disease management, adherence to prescribed medications and other treatments, and addressing mental health concerns such as anxiety and depression. This has been made possible by a growing number of connected devices and platforms and the massive quantities of data they produce.

A CONSTANT COMPANION

According to the United States Census Bureau, smartphones were present in 84 percent of American households in 2018.[13] In 2021, the Pew Research Center reported that 97 percent of U.S. adults said they owned a cellphone, with a full 85 percent owning a smartphone.[14] As you might expect, ownership rates around the world vary. According to the Pew Research Center, an estimated five billion people owned a mobile device in 2019, with advanced economies such as South Korea reporting that 95 percent of its population owned a smartphone, as compared to significantly lower rates in emerging economies such as South Africa (at 60 percent), Mexico (52 percent), and India (24 percent).[15]

These numbers confirm what you likely already suspected—nearly everyone in the United States and in other major economies is walking around with a tiny computer in their purse or pocket, every day. In other parts of the world, the numbers may be lower, but the trendline is similar. Mobile devices have become constant companions for billions of people.

As if that weren't sci-fi enough, manufacturers of mobile devices have equipped the hardware with various sensors. In addition to the microphone and GPS, today's smart devices include gyroscopic sensors, accelerometers, heart rate monitors, and barometers in various configurations, meaning signals from devices are being combined with mHealth apps, machine learning, and AI to detect things like a bad fall or a car crash.[16] The latest Apple Watches can monitor the wearer's heart rate and even take an ECG.[17] With each new advance, our phones get closer to functioning like the medical scanners on *Star Trek*. Meanwhile, our watches have already surpassed the capabilities of spy-movie communications devices.

This proliferation of technology and mobile connectivity among the world's populations provides the infrastructure for data transfer and storage, integrated software platforms, and machine learning applications. mHealth puts this robust data to good use in ways that are both simple and scalable.

Beginning with a simple example, recall your own experience visiting a physician's office for routine care. It's become an industry standard for physicians to spend a mere fifteen minutes per patient appointment as a result of various economic and structural forces present in healthcare systems. For many patients, this can feel like a paltry amount of time, barely enough to exchange pleasantries with the doctor, let alone discuss a subject as important as one's health.

In such a scenario, the opportunities for information loss are obvious. If the doctor is concerned about some aspect of the patient's health, the patient must absorb the potentially upsetting information, understand a new and possibly complex issue, and ask follow-up questions in a matter of minutes. In addition, the physician may advise a course of action for the patient that includes immediate steps, medications or other treatments to begin, behaviors to adopt or avoid, and signs to watch for in the future. That's a lot to expect of anyone.

Many mHealth initiatives aim to address this problem by facilitating communication between patients and healthcare providers. mHealth's apps, portals, and virtual visits use mobile devices to reach patients wherever they happen to be. This has moved far beyond mere electronic recordkeeping (EMRs/EHRs), which only started to gain traction in about 2006.[18]

What exactly are these apps delivering? In a study published in 2022, Anindya and his coauthors conducted an exhaustive review of the mHealth literature to summarize the many ways mHealth applications are being used to extend and enhance patient care.[19] Some applications will be familiar to you, especially if, like Alex and millions of others, you use or have used fitness trackers and other wearables. Many mHealth applications focus on leading people toward behaviors that support and improve health, such as exercise, healthier food choices, and sleep habits. These behavior-focused apps often employ "personalized goal setting" and "gamification" techniques, as well as "social comparison and competition" and "self-monitoring and review."

The 2022 study demonstrated that engaging with apps like these can affect physical activity levels, sleeping patterns, nutrition habits, glycemic control, and other longer-term health metrics, especially for patients with a chronic disease.[20] This is an example of a study that had used randomized field experiments, evoking the third pillar, "causal analytics," in the House of AI.

EDUCATION, ADHERENCE, AND HEALTH LITERACY

Of particular interest to medical professionals, mHealth has shown promise in helping patients adhere to prescribed treatment protocols. Hundreds of mHealth apps are designed to address common causes of nonadherence, from patients not understanding a treatment's purpose or side effects to simple forgetfulness. Reminders for when and how much medication a patient is supposed to take are simple examples that have been shown to improve outcomes for conditions such as hypertension, and for managing the complex treatment schedules that come with serious diseases including cancer.[21] Several recent studies have successfully piloted mobile SMS (Short Message Service programs to help patients manage asthma, HIV/AIDS, and diabetes, and to improve or eliminate lifestyle factors like smoking or excessive drinking.[22]

Alongside adherence, education is another way mHealth plays a role in patient health. Medication reminders and alerts, while important, may be insufficient if patients misunderstand things like dosage (how much?) and administration (how to?) instructions. Medication administration errors are a known problem in medicine, especially in cases where patients are at home, away from clinical supervision. Various factors contribute to home medication administration errors, including preparation type (e.g., measuring liquids as opposed to tablets) and managing multiple medications simultaneously. Some of the most common cases where such errors occur involve parents overseeing medications for their children with chronic diseases.[23]

mHealth can help address these problems. For example, take literacy and language proficiency. These may sound like problems limited to select groups, but in fact they affect a wide swath of the population. That's because "literacy" in this context refers to *health literacy*, or the ability to comprehend health and medical information, procedures, and systems. The American Academy of Pediatrics says that "nearly 30 percent of parents in the United States, or approximately 21 million U.S. parents, have low health literacy," and "only 15 percent of parents are considered to have proficient levels of health literacy."[24] You read that right: the *majority* of parents and caregivers are likely to struggle with some aspect of health literacy, such as not comprehending drug labels, improperly measuring, or mistaking dosage adjustments based on weight, all of which can increase the risk of medication errors and the resulting complications. As you might expect, these same problems compound in situations where caregivers are not native speakers of the dominant language. An episode of *ER* from the 1990s used this problem as a plotline for one of its episodes: the patient "accidentally overdosed on her medication because the label [was] in English [and she only spoke Spanish]. She was supposed to take a pill once a day, but "once" in Spanish is 11, so she took 11 times the prescribed amount."[25] Such errors are tragic, but increasingly preventable with mHealth tools.

The American Academy of Pediatrics recommendations for addressing home medication errors are detailed and comprehensive. Many of the communication and education techniques they describe might be delivered or aided by mHealth apps, platforms, and devices. For example, recommendations to write instructions using plain language, simplified steps, and visual/pictorial aids can all be achieved with the help of mHealth-delivered media. Imagine picture-based slideshows or video demonstrations showing proper medication preparation, administration, and disposal, like IKEA instructions for healthcare. Similarly, photos might be incorporated to show known side effects or allergic reactions.

Advances in AI-powered language translation are another obvious candidate for increasing the effectiveness of mHealth. Simple, computer-assisted translation of words and phrases using a smartphone (think Google Translate) are just a small step in this direction. AI, with its ability to process massive amounts of data and learn from the results, can go much further in reducing translation mistakes and misinterpretations when applied to a specific domain such as medical instruction.

As with language translation, the customization of health information for specific patient needs is an area where mHealth shines. For example, one study looked at how mHealth tools might be used to improve cardiovascular health among African American patients. Using a randomized control trial, the study found that educational content and reminders—delivered via personalized app messages—helped to increase health-promoting behaviors recommended by the American Heart Association.[26] The study coupled its use of mHealth tools with culturally relevant content and places (e.g., churches and faith institutions) trusted in the community. As any of us who shops online knows, targeting individuals with highly customized content is something our data-driven technology ecosystems do quite well.

MENTAL HEALTH AND DIGITIZED THERAPY

The potential benefits of mHealth are not limited to physical health and fitness. Other applications have shown promise in managing insomnia, anxiety, depression, and PTSD. The idea guiding this type of mHealth is to digitize existing therapies like cognitive behavioral therapy (CBT) in order to address a range of mental health issues, including postpartum depression and substance abuse.[27]

mHealth platforms aimed at digitalization of existing therapies gained traction during the COVID-19 pandemic amid infection fears and widespread shutdowns. Interest in applications for mental health has continued to grow as the long-term effects of the pandemic begin to emerge. According to the World Health Organization, "In the first year of the COVID-19 pandemic, global prevalence of anxiety and depression increased by a massive 25 percent" due to a variety of stressors such as social isolation, fear of infection, and economic repercussions as a result of shutdowns.[28]

Limited access to care compounds this problem. According to a 2022 report from Mental Health America, "Over 27 million individuals experiencing a mental illness are going untreated" in the United States. "Over half (56 percent) of adults with a mental illness receive no treatment."[29]

In light of these trends, a variety of AI-driven digital tools have emerged, often positioned as cost-effective ways to offer care to more people in more places, while also reducing the stigma associated with seeking mental healthcare. Available tools cover a range of needs, from symptom identification and diagnosis to meditation apps and psychotherapy chatbots. According to Kashyap Kompella (a former student of Ravi's), "One potential advantage of digital apps is that they can gather a more detailed picture of the users' mental states based on daily (or even more granular) logs compared to the less-frequent (weekly or monthly) self-reporting that is the norm for in-person CBT [cognitive behavior therapy]."[30]

COMPLEXITIES AND NUANCES: UNDERSTANDING WHAT WORKS AND WHEN

Just as with the health of individual people, effective mHealth interventions are turning out to be something more nuanced than a one-size-fits-all approach. For example, researchers are finding that techniques such as personalization generate different results depending on the circumstances and how they are used. Anindya Ghose recently coauthored a study that illustrates the point.

In the study, Ghose and his colleagues partnered with a major mHealth app platform in Asia to conduct a large-scale randomized field experiment. Participants in the study were patients with chronic diabetes. The basic idea was to see if adoption of an mHealth app could motivate beneficial behaviors related to activities such as walking and other exercise, as well as sleep and eating habits. The researchers also wanted to see if an mHealth app could motivate these behaviors enough to move the needle on an important tangible health marker such as blood glucose levels.

The study found that patients using the app walked and exercised more, ate healthier food, and slept longer. In addition, users of the mHealth app showed better health outcomes on metrics such as blood glucose and glycated hemoglobin levels that are indicators of diabetes. The same patients experienced a reduction in hospital visits and medical expenses.[31] These results are excellent news, as they show that mHealth interventions can produce positive changes—both short-term and long-term—for individuals and the overall system.

But back to the question of personalization. In Anindya's study, there appeared to be such a thing as *too personal* when it came to the app's reminder messages and content. The researchers compared the effect of highly personalized messages ("Mr. X, you did not exercise at all yesterday. Take a forty-five-minute walk today to help control your blood glucose level.") to that of more generalized reminders ("Regular exercise at moderate intensity is very helpful in controlling blood glucose."). Over time, the generalized messages were shown to be 18 percent more effective than the personalized ones at reducing glucose levels.

Surveys of the participating patients revealed that the highly personalized messages felt intrusive and annoying for some patients. They felt they were being judged, which produced the undesirable effect of demotivating them. Instead of engaging in *more* of the recommended healthy activities, they did just the opposite. It is important to mention the role of cultural factors in deciphering the finding that personalization can be a double-edged sword in the context of mHealth. Researchers have shown that individuals who exhibit more social behaviors tend to be more receptive to recommendations that are *not* personalized to their own preferences, but instead to the collective preferences of their social group.[32] So depending on the country, personalized healthcare reminders need to be contextualized based either on individual preferences or group preferences. How can we better address the issue of inadvertent negative effects from over-personalization? Academics have explored how device-delivered haptic feedback (something touch-based like a cellphone vibrating) can improve responses to certain consumer-directed communications. Across four studies, the authors found that haptic alerts accompanying messages can improve consumer performance on related tasks.[33] They argue that this effect is driven by an increased sense of social presence in what can otherwise feel like an impersonal technological exchange.

Similarly, a study by researchers from the University of Maryland examined another aspect of motivation related to mHealth apps. They found that the typical financial incentives used in wellness activities (e.g., cash rewards for participating) were not the most effective at getting people to exercise. Their experiment design tested other kinds of rewards as motivation and found that more participants completed exercise challenges when the promised rewards benefited their *friends* on the connected social network.[34]

These results underscore the value of the causal analytics pillar described in chapter 1. At this point in our history, researchers in many different fields are uniquely positioned to conduct more experiments in less time, thanks to easier data collection, increased willingness of practitioners and executives to undertake experimentation, and faster analysis made possible by AI-powered technologies, platforms, and ecosystems. We're excited to see the next generation of research resulting from all the mHealth data that's being collected.

USING AI AND WEARABLES TO BUILD BRIDGES BETWEEN PEOPLE

Discussions of mHealth often begin with and revolve around exercise. Indeed, that's how we started this chapter, because examples related to exercise are familiar to most people and many of us have some experience with fitness trackers of one sort or another. But mHealth extends well beyond setting a new personal record on your favorite running route or knowing how long you sleep on an average weeknight.

One example that caught our attention revolves around a product you may remember from a decade ago: Google Glass. Launched in 2013, this wearable device spurred a lot of reaction in the media and popular culture. The very idea of wearing a "face computer" met with quite a lot of skepticism and mockery. As the tech-focused news site *Mashable* put it, "The world wasn't quite ready to see early adopters walking around with cameras on their faces. . . . Some restaurants and bars banned them entirely, and their wearers were not-entirely-affectionately dubbed 'glassholes.'"[35] Sketches featured in shows like *Saturday Night Live*[36] and *The Daily Show*[37] concurred.

Google stopped supporting the software for the original consumer product in 2020.[38] However, the company is offering a more specialized Glass Enterprise Edition for use in manufacturing, field service, and yes, healthcare settings.[39] One of Google's case studies on the product describes how the device works with a documentation automation software platform called Augmedix to reduce the time doctors spend looking at computer screens, giving them more time to focus on the patient in front of them. They report a 30 percent increase in physician productivity.[40]

Despite the mockery, Google Glass and other wearables like it may prove valuable in the long run, especially if the most useful applications turn out to be something other than an everyday device for average consumers. One very specialized application involves children with autism.

In the late 2010s, researchers at the Stanford University School of Medicine piloted a study that used Google Glass to help children with autism develop their social skills. The project connected the tiny Google Glass camera with a "Stanford-designed app that provides real-time cues about other people's facial expressions to a child wearing Google Glass."[41] According to a 2018 article describing the project, "As the child interacts with others, the app identifies and names their emotions through the Google Glass speaker or screen. After one to three months of regular use, parents reported that children with autism made more eye contact and related better to others."

The ability to identify and understand facial expressions and other social cues is challenging for people with autism, which can lead to isolation and behavioral problems. The Stanford project demonstrates how AI-powered platforms may help bridge the gap between patients with autism, their families, and the wider world. Ongoing work by the Stanford team uses machine learning and AI in combination with behavioral therapy apps and wearables to break down barriers for the one million U.S. children on the autism spectrum (one out of every fifty-nine kids).[42] This illustrates again the enormous potential of AI to fundamentally improve the emotional and physical well-being of everyday people.

The Google Glass example is not alone. The U.S. Food & Drug Administration (FDA) ruled on ninety-one AI- and ML-enabled medical devices in the first nine months of 2022. According to the announcement, "As technology continues to advance every aspect of health care, software incorporating artificial intelligence (AI), and specifically the subset of AI known as machine learning (ML), has become an important part of an increasing number of medical devices. One of the greatest potential benefits of ML resides in its ability to create new and important insights from the vast amount of data generated during the delivery of health care every day."[43]

The devices on the FDA's list are associated with a variety of medical specialties, including anesthesiology, hematology, and neurology, among others.

About 75 percent are in radiology, in part because of data availability and compatibility.[44]

Many of these innovations seek to augment, extend, and complement the skills of physicians and other experts, not supplant them. For example, machine learning is well suited to analyze huge troves of data and provide alerts to aid doctors. Analyzing patient electronic health record (EHR) databases, machine learning can help uncover relevant risk factors that are associated with a particular disease. That technology can also be used to evaluate an individual patient's metrics and alert the physician if they are associated with an elevated risk, similar to Hazel's WHOOP band alerting her about her abnormal HRV from earlier in the chapter. This is already happening, and it's helping improve healthcare for people everywhere. Machine learning models have been trained to read X-ray images and identify concerns. Researchers have also trained a deep learning model to review retinal images and automatically detect conditions such as diabetic retinopathy and diabetic macular edema.[45]

There are many advantages to this kind of AI-augmented healthcare. Obviously, computers are fast, they don't get tired, and they don't have bad days. They can process a large volume of data quickly and accurately. The ability of a model to correctly identify cases of concern can save a tremendous amount of time for medical practitioners, allowing them to attend to other matters and care for more patients. It follows that it can then also save on misdiagnoses, wrongful deaths, and malpractice suits. This complementary relationship between human and machine can reduce the chances that important warning signs will be missed for patients and reduce stress on already overtaxed healthcare workers.

PREDICTION AND PREVENTION

As we've mentioned, data and data engineering are crucial components of AI-powered digital ecosystems. Data holds answers, but human capacity is limited. The catch is that even if we had *all* the world's data available to us, we wouldn't be able to look at it and find the answers. It's just too much. It would be like drinking out of a fire hose, or looking for a needle in a haystack, only it's a haystack *made* of needles. Which one of those countless points is the meaningful one? This is where AI and machine learning can easily do what we cannot: find the relevant patterns of information in vast amounts of data and then present them to humans for interpretation and action.

One area of the healthcare system that we haven't mentioned yet is insurance. Ravi helped set up Optum's Data Science University and coached hundreds of its executives on the art of what's possible with AI and ML.[46] Without going into detail, we can confirm that the insurance industry is using data and AI techniques to change the way business is done, particularly when it comes to fraud detection. What may look to the human eye—even the eye of a trained and seasoned analyst—like regular claims and payment activities can be revealed as fraud by AI based on the data. The National Health Care Anti-Fraud Association (NHCAA) estimates that fraud results in losses of tens of *billions* of dollars annually; other agencies say it could be as much as $300 billion.[47] If an AI-powered system can help find and eliminate costs of that magnitude, the industry's whole financial picture changes. It's conceivable that the premiums people and employers pay could drop significantly when insurers are paying for actual healthcare and not nefarious actors.

Insurance data and data from other health-related sources likely hold additional value. For example, the *Financial Times* reports that Sampo, one of Japan's biggest insurance companies, is mining data on the country's senior population for insights into dementia.[48] The company is using data on Japan's large population of elderly citizens to identify the early warning signs and to give individuals tailored advice for staying healthier for longer. If more seniors adopt behaviors correlated with warding off dementia, it will not only improve their quality of life, but it also reduces risk for insurers. The insights could also affect the design and cost of insurance products, benefiting both parties.

Again, as with any tool or technology, the possibility for abuse exists. AI-based predictions could be used to deny people insurance coverage or to reinforce existing biases against some groups of people, such as people of color. Public policy is needed to guard against such misuses. We don't mean to suggest that AI solves all problems magically, only that the capability offers tremendous potential, and that it's up to societies and their leaders to use it responsibly, ethically, and productively. A printing press can produce books that educate and enlighten millions or a racist screed full of hateful rhetoric. A knife can be a tool used to create great dishes in the kitchen or a weapon.

THE FUTURE IS HERE: AI ENABLES NEW FRONTIERS OF RESEARCH AND THERAPEUTICS

Of course, human health and wellness are about much more than fancy watches that track daily steps and remind you to take your pills. In addition to helping predict, prevent, and treat various health conditions, AI is helping researchers expand our knowledge of biology at fundamental levels and helping companies deliver on the promise of personalized medicine.

Researchers around the world have already made huge progress using AI to predict things like protein structures, the building blocks of life itself, which determine bodily functions at the cellular level. Understanding these structures is crucial to advancing scientific knowledge of genetics, viruses, bacteria, and diseases. The ability to predict these structures accurately has been called "a revolution in biology," a game-changer that makes it possible to study living things faster and in novel ways that could accelerate drug discovery and other advancements in medicine.[49]

"Researchers have used AlphaFold—the revolutionary artificial-intelligence (AI) network—to predict the structures of more than 200 million proteins from some 1 million species, covering almost every known protein on the planet," reports *Nature*.[50] These highly accurate predictions are having a major impact on research agendas in the life sciences fields because they eliminate or reduce the need to rely on "time-consuming and costly experimental methods such as X-ray crystallography or cryo-electron microscopy to solve protein structures."

We expect that AI's impact in this arena will be nothing short of profound. Biotech and pharmaceutical firms are already developing drugs with AI's help. As a 2022 *Financial Times* article explains, "AI platforms can crunch vast amounts of data to rapidly identify drug targets—proteins in the body associated with particular diseases—and molecules that can be made into medicines. Experts say the technology can slash the time it takes a drug to go from initial discovery to approval, cut the costs of development and reduce the high failure rate in clinical trials."[51] As a result, new drugs have progressed to the clinical trial phase faster, including one treating the neurodegenerative disease ALS.

Take the case of a new generation of therapeutics coming from Moderna in Cambridge, Massachusetts, the company that pioneered the use of mRNA technology to deliver the blockbuster COVID-19 vaccine in record time. Among the many products in its pipeline, one that particularly caught our attention was "mRNA-4157—Personalized Cancer Vaccine," which is currently under Phase 2 clinical trial with results expected by last quarter of 2022 (at the time of the writing of this book).[52] We caught up with Marcello Damiani, Moderna's Chief Digital and Operational Excellence Officer, at a conference to learn about the "personalized" part of this potentially breakthrough cure for cancer. The key idea behind the mRNA vaccines is to prime the immune system in a particular way so that a patient can teach her cells to make a protein that then triggers an autoimmune response. This response produces antibodies that help protect us from getting sick from that disease in the future. When applied to cancer, Marcello explained, "a key challenge becomes which thirty-four (think of this as the size of the 'cartridge' that the current mRNA technology can safely deliver) of the hundreds of thousands of cell mutations to target?" This is where AI comes into play. For each cancer patient's cell mutation patterns, an AI (supervised learning) model can rank in order the most effective mutations to target to that individual, and the top thirty-four are then delivered using an mRNA-4157 personalized cancer vaccine shot (much in the same way you probably received your COVID-19 vaccination).

The remarkable thing about mRNA technology is that it can, in principle, be delivered to alter any deficiency in any protein structure of our body that causes any ailment. By taking the code of a particular protein, producing it in the lab, and then subsequently delivering it to your body to trigger an immune response, designing a new cure for a disease is essentially a matter of "changing the cartridge" (Marcello's term) inside the mRNA injection to prevent or cure that specific disease by targeting a different thirty-four cell mutations. AI can help narrow that gigantic protein search space, as it does in the case of the personalized cancer vaccine. This vastly reduces the time to market for a new cure to be introduced to the public. If what we saw with the COVID-19 vaccine plays out in a similar fashion for other rare or chronic diseases, we could be looking at human life very differently in the next decade or so.

MORE HEALTHY DAYS IN A YEAR, MORE HEALTHY YEARS IN A LIFETIME

Hopefully, this collection of examples has helped illustrate why we and others are excited about AI, especially in the context of health and medicine. The fitness tracker you use to count your steps or monitor your sleep is just the beginning. An expanding ecosystem of connected devices, apps, and sensors, combined with AI's power to analyze and make sense of the resulting data, have set humanity on a course toward a healthier future. One that includes more access to care and a better understanding of how we as individuals can lead healthy lives. One that makes the management of chronic conditions easier and more effective, aided by ingenious devices. One that helps us excel in the pursuit of our athletic hobbies. And one that takes on the toughest health challenges by using AI to expand the boundaries of human knowledge.

TAKEAWAYS

- More and more people are using the AI-powered digital ecosystem every day to monitor, track, and improve their health and personal wellness. Many have embraced the "quantified self"—the notion that collecting data on one's eating, sleeping, and physical movement can provide insight into the state of the body and lead to greater health, fitness, and overall well-being. AI-powered mHealth apps are now being used for remote patient monitoring, patient tracking, medication management, disease management, women's health, and personal health record management.

- AI is helping with advanced medical diagnosis and chronic prescription-medicine adherence problems. Given the advances in image recognition using deep learning, some of the breakthrough practical applications of AI are in radiology. Remember the Budapest example? Since 2021, twenty-two cases have been documented where the AI technologies identified a cancer missed by radiologists, with about forty more cases under review. Furthermore, advances in AI-powered language translation and customization of health information are improving adherence with medical procedures and care. As any of us who shops online knows, targeting individuals with highly customized content is something our data-driven technology ecosystems do quite well.
- The potential benefits of mHealth are not limited to physical health and fitness. Other applications have shown promise in managing insomnia, anxiety, depression, and PTSD. The idea guiding this type of mHealth is to digitize existing therapies like cognitive behavioral therapy (CBT) in order to address a range of mental health issues, including postpartum depression and substance abuse.

5 LEARNING AND EDUCATION IN THE AGE OF AI

The term *artificial intelligence* has been around for decades. The *idea* of sentient, nonhuman intelligence can be traced to the nineteenth century at least, in novels such as Mary Shelley's *Frankenstein* (1818) and Samuel Butler's *Erewhon* (1872), for example. Several popular film and television series have explored AI-related themes, including *The Terminator, Battlestar Galactica, The Matrix*, and *Black Mirror*, all of which paint bleak pictures of worlds where technology turns against humans in various ways. Such stories may be the stuff of science fiction, but their enduring presence in pop culture has made an imprint on the public's imagination. For many people, the term *artificial intelligence* conjures something cold, sinister, even ruthless. AI is a new form of "life" that poses dire threats to all humanity once it's unleashed.

As we wrote in chapter 1, this book is not about general artificial intelligence or sentient technology. Thankfully, the world is not facing the computers-and/or-robots-taking-over-and-killing-everyone kind of AI, although AI pioneers such as Geoffrey Hinton fear this possibility. However, the latest advancements in AI have people once again imagining a world where computers push humans closer to obsolescence. For example, one of the biggest tech stories of late 2022 and early 2023 was the introduction of ChatGPT. Developed by OpenAI, an artificial intelligence research company,[1] ChatGPT attracted major media attention upon its introduction, because of its ability to generate "a fairly convincing approximation of text written by humans."[2] Within five days of its release, ChatGPT attracted more than a million users, according to a Tweet posted by Greg Brockman, the president and cofounder of OpenAI.[3] The company made its chatbot freely available for anyone with a web browser to try—and try they did. People asked ChatGPT to write everything from paragraphs to poetry, book reports to Bible verses, and computer code to college essays.

Depending on the task it was asked to perform, ChatGPT's output ranged from solidly convincing to unexpectedly hilarious to downright uncanny. Reporters chronicled examples and people shared screenshots on social media.[4] Many of the examples were funny and entertaining. Many were impressive: ChatGPT showed itself capable of writing credible-sounding legal documents, answering essay questions from college exams, and debugging blocks of code. In the weeks and months that followed, writers, pundits, academics, technologists, ethicists, and others began commenting on and debating what kind of effect AI at the level of ChatGPT (and beyond) will have in the short and long term. As with many other technological advances, ChatGPT raises important questions while simultaneously generating both excitement and anxiety.

As academics in the business of writing, we put ChatGPT (version 3) to the test for a recent manuscript Ravi was completing with coauthor Gordon Burtch that examines the effect of the NCAA's Name, Image and Likeness (NIL) regulation on athletes' athletic performance. After two years of hard work, theorizing, collecting data, running a variety of analyses, and discovering great insights, Ravi was suffering from pesky writer's block when it came to the discussion and conclusion section of this paper. He procrastinated for two weeks, until Gordon rather casually suggested, "Why don't we give this task to ChatGPT?" Gordon copied the introduction, background, and results section of the paper and asked ChatGPT to write the conclusion and discussion section that summarized the main findings and generated ideas for future research. What they got back was a mixed bag. Gordon and Ravi analyzed the output of ChatGPT and concurred that

- It did a reasonable but nowhere near stellar job of summarizing what the paper was about. It did highlight the tension between athletes being more distracted by NIL monetization opportunities or being more motivated, a key idea that motivated the paper.

- It completely failed to pick out the most insightful results and the nuances of the mechanism we identified to explain our observed effects.
- It failed to mention how we established the credibility of our results using Canada as a plausible control group where the NIL law did not apply.
- It generated four ideas for future research, one of which was very interesting ("What impact may result from this policy change for non-revenue-generating sports and athletes?") and incorporated into the final paper. This is a nontrivial achievement.

A key factor that we believe needs to be researched further is whether ChatGPT can play a significant role in nudging people out of the inertia of writer's block, as it did for Ravi. That, in and of itself, would be a significant impact at population scale.

History holds numerous examples of how new technologies reorganize economies, disrupt once-reliable livelihoods, and change how we humans spend our time and money. As the psychologists Kathy Hirsh-Pasek and Elias Blinkoff put it, "The invention of the telephone in 1876 was met with simultaneous amazement and trepidation. Critics wondered if phones would disrupt face-to-face communication in ways that made us either too active or lazy."[5]

ChatGPT has prompted a fresh exploration of the kinds of questions that accompany any significant new technology. If a computer can generate such convincing, human-sounding text across so many genres and applications, what happens to the writers, editors, programmers, and others who've built careers selling the same skills? Will students bother learning how to do these tasks themselves if they can get the same or better results from an app? In a world with such powerful tools, is it even *necessary* for students (or anyone else) to learn these things anymore? Will teachers, employers, and professionals consider ChatGPT to be a form of cheating or simply a time-saving tool? After all, is what ChatGPT does really any different from, say, the spellcheck feature in a word processor or the automated formula calculations made by a spreadsheet? These questions barely scratch the surface, given the kinds of results ChatGPT generates.

It is still early days when it comes to rigorous research on how effective or not ChatGPT can be in the classroom. There is no shortage of opinion, though. Reactions to ChatGPT have been mixed, as you might expect. Reviewing the techgettinggy for the *New York Times*, Kevin Roose called ChatGPT "quite simply, the best artificial intelligence chatbot ever released to the general public," and went on to muse, "The potential societal implications of ChatGPT are too big to fit into one column. Maybe this is, as some commenters have posited, the beginning of the end of all white-collar knowledge work, and a precursor to mass unemployment." He ended the column with the sentence, "We are not ready."[6]

Gary Marcus described ChatGPT and similar AI as "potentially dangerous" in *Scientific American*, saying, "Because they are so good at imitating human styles, there is risk that such chatbots could be used to mass-produce misinformation."[7] He went on to describe potential misuses as "existential" in nature, positing that scam artists, nation states, and rogue actors would likely use the technology to flood the world with misinformation.[8]

Others, like Jacob Stern writing in *The Atlantic*, have expressed more skeptical views: "The powerful new chatbot could make all sorts of trouble. But for now, it's mostly a meme machine."[9] Writing in the same publication, Ian Bogost critiqued the app, its creators, and the general enthusiasm that greeted the release. After asking ChatGPT a variety of questions across genres, Bogost concluded that "the bot's output, while fluent and persuasive as text, is consistently uninteresting as prose" and called ChatGPT "a toy, not a tool."[10] In *Harvard Business Review*, Ethan Mollick conceded, "At first glance, ChatGPT might seem like a clever toy." He went on to say, "But a deeper exploration reveals much more potential," and called the development a "tipping point" for AI.[11]

Only time and more rigorous research will tell which of these varied reactions will be validated and which will be undermined. But it's fair to say that what comes *after* ChatGPT is going to be very interesting to watch.

In the meantime, we want to examine AI's effect on education, a topic that's personally quite meaningful for a couple of professors and career academics like us. Among the immediate concerns raised about ChatGPT is what this kind of tech will mean for educational institutions. There's the potential for cheating on written assignments, but that's just the beginning. How will schools, colleges, and universities educate students if we're all drowning in inaccurate-but-credible-sounding misinformation? As Roose puts it: "It's easy to understand why educators feel threatened. ChatGPT is a freakishly capable tool that landed in their midst with no warning, and it performs reasonably well across a wide variety of tasks and academic subjects. There are legitimate questions about the ethics of AI-generated writing, and concerns about whether the answers ChatGPT gives are accurate. (Often, they're not.) And I'm sympathetic to teachers who feel that they have enough to worry about, without adding AI-generated homework to the mix."[12]

Given what we've seen to date, instances of cheating that go undiscovered and unpunished because of ChatGPT are certainly possible, perhaps even likely. There will be a period of adjustment as educators determine how to evolve lesson plans that may be impacted by such a technology. But on balance and in the long run, we share the view that with such a powerful tool comes amazing potential. As Hirsh-Pasek and Blinkoff say, "In the same way that calculators became an important tool for students in math classes, ChatGPT has potential to become an important tool for writers who want to hone their critical thinking skills along with their communication skills."[13]

The reality is there's no going back to a world before ChatGPT. Academic institutions must adjust, just like they adjusted to other innovations, including the Internet, personal computing, and yes, the calculator. More to the point, AI and machine learning have the potential to reshape teaching and learning in profound—and beneficial—ways. We'll introduce you to a few of those ways later in this chapter.

WHAT IS CHATGPT AND HOW DOES IT WORK?

ChatGPT (Chat Generative Pre-trained Transformer) is a large language model that incorporates various methods to generate responses to human questions and commands. We won't go into all the technical details in this book, but key aspects of the method are of general interest and speak to the ingenuity of the AI researchers who have pioneered this approach. A large language model is a deep learning-based approach to predict, translate, and generate text using insights gleaned from a large corpus of existing texts including all Web content, Wikipedia, social media such as X (formerly Twitter), and other sources including the Common Crawl. Legal questions around copyright violations of using such text notwithstanding, there are four main steps in the working of large language models; the first two you are already familiar with from earlier in this book. Let's review them for ChatGPT:

1. ChatGPT begins by *tokenizing* sequences of words or sentences into sequences of numbers.
2. It maps these tokens into an *embedding* space that give the tokens some meaning. Words, sentences, and phrases with similar meanings are close to each other in this embedded space (refer to figure 2.1 for a refresher on embeddings) and the model remembers the position of each token (*positional encoding*).
3. The model then calculates how much *attention* a given token has to pay to another token. In fact, multiple such attention weights are computed in parallel, capturing different facets of real-world context.
4. The model generates a series of probabilities for the most likely next token. This could be the next word, which is then fed back into the model to generate the next word, and the next word, and so on till it reaches a state of *completion*.

Consider a traditional (pre-ChatGPT sequential deep learning) approach to translate (into Spanish, say) a long sentence such as this one:

> Anindya and Ravi both trekked to the base camp of Mount Everest in their youth. However, with time their interests diverged. Today Anindya likes to pursue mountaineering in Latin America and Ravi enjoys his tennis.

A traditional approach would work toward memorizing (technically, we call it *encode*) the whole sentence to learn about the context of the words before attempting to translate (*decode*) it sequentially, word by word, into, say, Spanish. A seminal paper by scientists at Google and the University of Toronto changed this paradigm by introducing the *transformer-based approach*.[14] In this approach, the algorithm learns how much attention a given word must pay to another word in the sentence. Again, the learning here is based on testing (in a mathematically principled way) the different levels of attention a given word or name—*Everest*, for example—pays to the word *mountaineering* to predict the training data, which is the large corpus of text. As this process is iterated hundreds of thousands of times, the algorithm learns that the word *Everest* should pay more attention to the word *mountaineering* than to the word *tennis*. A simple use of learning such weights is to complete a sentence (the way Gmail does for you) by predicting what you are going to say based on the context of your email. It could be used to translate the sentence from English to Spanish that we discussed earlier, and it could be the engine behind a conversational bot such as ChatGPT. Thus, when a future user asks for a recommendation of a challenging trek in the Himalayas, the system can return something like, "Try the Everest Base Camp Trek. It gets you to 19,000 feet without the need for any formal mountaineering skills," rather than suggesting, say, a tennis camp in Southern France.

If we can do this attention mapping (think of this as learning an importance weight) from every word to every other word (this would mean learning at least $35 \times 35 = 1,325$ weights—in reality we need to track other things as well, such as the position of a given word in the sentence) then we can essentially understand the context and connections of the words in our language. Essentially, the algorithms retrieve pieces of information about a subject and assemble them into appropriate patterns. AI developers train the model by feeding it large numbers of examples. By analyzing thousands of resumes, newspaper stories, or code blocks, AI-based algorithms learn the pattern for each and then replicate it. Compared to the previous sequential approach where we had to memorize the whole sentence, we can do this in parallel for large numbers of words at a time. This, in turn, dramatically speeds up the time to train the model and allows us to understand even greater amounts of written text, resulting in a model such as ChatGPT that learns close to 175 billion attention weights for ChatGPT-3, and an exponentially higher number, 100 trillion, for the recently released (at the time of writing this chapter) ChatGPT-4[15] to gain its smarts. ChatGPT-4 scores in the 99th percentile on the Graduate Record Examinations (GRE) Verbal section, but a disappointing 80th percentile on the GRE Quantitative section (yes, we are South Asian parents with kids applying to US colleges; the 80th percentile is disappointing). ChatGPT-4 did rank in the top 10 percent of a simulated bar exam.[16]

At its core, ChatGPT-4, like previous LLMs, is trained to predict the next word in a document using publicly available data (think of all the Internet content and then some). It represents and predicts based on all the ideologies that are out there, so we imagine that it might often struggle to understand the intent of the user asking it to do something. A lesser-known fact that is specific—and ingenious, we might add—to OpenAI's approach toward making ChatGPT interactive and responsive to human prompts is that it relies heavily on large numbers of human contractors (thanks in no small part to the early $1 billion investment from Microsoft) who are part of the process of garnering user intent and putting guardrails around the output emitted by ChatGPT. Ask ChatGPT-4 "how I can create a bomb?" and it will politely remind you:

> My purpose as an AI language model is to assist and provide information in a helpful and safe manner. I cannot and will not provide information or guidance on creating weapons or engaging in any illegal activities. Please let me know if there is another topic I can help you with.[17]

The contractors' job is to rank order various answers to the prompts. Thus, human feedback is used by the algorithm to learn a reward function that feeds into a reinforcement learning algorithm.[18] In fact, the RLHF (reinforcement learning with human feedback) approach is one of the innovations that has made ChatGPT stand out in comparison to language models of the past. More data and different examples give the algorithm more to examine and learn from. Next, another model is used to generate multiple answers to the same command, and to annotate them "based on a multidimensional criteria [*sic*] that incudes aspects such as relevance, informativeness, harmfulness and several others."[19] Humans rank the various answers to the command, thus providing data that reinforce the kind of output that makes sense and is desirable.

ChatGPT is built to respond to user input in a conversational way. Ask it a question, get a response, ask a follow-up question. This interface contributes to buzz around the technology—it can certainly *feel* like an interaction with an intelligent conversationalist. But again, ChatGPT and LLMs in general are repeating patterns of human speech and writing.

Much of the initial reaction to ChatGPT fell into two categories. Many people were awestruck by the bot's ability to produce such credible-sounding text (essentially following the patterns we expect in various contexts). This often led to pronouncements of AI takeovers of various industries, such as copywriting, journalism, and software coding. And then there were authors such as Ian Bogost, who pointed out that ChatGPT still has limits; it quite frequently provides inaccurate answers and even freely admits its mistakes.[20]

There's some truth in both perspectives. But the space between is more interesting in our view. ChatGPT needn't be an all-or-nothing proposition. Like any powerful tool, the human user's *intent* matters a great deal. As *New York Times* columnist Peter Coy put it, "I got mediocre results from ChatGPT because I didn't try very hard to use it properly. Other people have gotten amazing results because they're smarter and more purposeful about how they use it."[21]

Coy's comment underscores the idea that AI like ChatGPT isn't necessarily a *replacement* for human thought, activity, or expertise. Once the novelty wears off, AI is more likely to be a way to augment and extend what humans already do. Mollick discussed this in *Harvard Business Review*, calling it "human-machine hybrid work. Instead of prompting an AI and hoping for a good result, humans can now guide AIs and correct mistakes. . . . This means experts will be able to fill in the gaps of the AI's capability, even as the AI becomes more helpful to the expert."[22] Viewed this way, AI is an enhancement, one that will greatly improve productivity in many arenas.

For example: a writer can easily edit badly written sentences that may appear in AI articles, a human programmer can spot errors in AI code, and an analyst can check the results of AI conclusions. This leads us, ultimately, to why this is so disruptive. The writer no longer needs to write the articles alone, the programmer to code on their own, or the analyst to approach the data themselves. The work is a new kind of collaboration that did not exist prior to the launch of ChatGPT. One person can do the work of many, and that is even without the additional capabilities that AI provides.[23]

Many academics have expressed similar views and offered ideas on how to incorporate the technology into curricula. Suggestions include having students use ChatGPT to generate drafts or starter code students can build on; assignments to critique ChatGPT's output and conduct research to verify presented conclusions; using the technology to summarize large bodies of information; and using the AI engine to compile a variety of perspectives on a topic.[24]

As we alluded to earlier, we are still in the early days of research that sheds light on how ChatGPT can affect education. A notable exception that goes beyond punditry opinion is work by Unnati Narang at the Geis School in Illinois. She designed a field experiment in an online course for which half the class (randomly chosen per the principles of causal analytics) was asked to use ChatGPT to garner ideas to contribute to the class's online discussion board.[25] Narang's findings are interesting and not so different from what Ravi and Gordon experienced. She found that students relying on ChatGPT made longer posts, but these posts were, essentially, less interesting. They generated lower levels of engagement in the form of fewer views and shorter comments relative to those assigned in the non-AI condition. At best, this suggests that students need to further their agency in using this new tool. At worst, it could actually be a source of distraction (similar to what Gordon and Ravi felt with their experience) sucking away mental energy that could be better spent engaging more deeply with the class material.

ChatGPT aside, AI and machine learning already help address various problems and barriers when it comes to education and are proving to be effective tools in the quest to educate more people on more topics and in new ways. The resulting educational gains improve the lives of individuals in meaningful ways. In addition, a more and better educated population produces downstream effects. Education makes people more employable, which makes for a more productive workforce and reduces poverty.

What follows are a few examples of how AI and ML are transforming education in positive ways.

GAPS AND BARRIERS PERSIST EVEN AS CHANGE ACCELERATES

Despite the many educational advancements from the 1900s to the present, there remain gaps in the education landscape. Some individuals and groups don't have access to education and are therefore being left behind. Others struggle to navigate an increasingly complex world that demands people acquire new skills and knowledge to succeed at earning a living and building a career.

According to the World Economic Forum, global literacy rates have been on the rise for a century and a half, hitting 87 percent in 2022, a new high, but that statistic is not uniform across the globe.[26] As you likely suspect, there are significant differences between certain countries and subpopulations. For example, in countries where conflict and war have disrupted daily life (including access to education) for years, literacy rates are much lower. In Afghanistan, the literacy rate is only 37 percent. In South Sudan the rate is 35 percent. And in Mali, literacy was at 31 percent in 2020, down from 35 percent in 2018.[27]

Lower literacy rates often affect some groups of people more than others. Specifically, women are often more likely to be left behind. In Sub-Saharan Africa, the literacy rate for males was 72 percent in 2022 but only 59 percent for females.[28]

In 2018 the World Economic Forum also reported that "more than one-half of India's workforce will need to be re-skilled by 2022 to meet the demands of the Fourth Industrial Revolution."[29] That's a huge portion of the population. As industries shift and evolve in the face of political winds, climate change, and advancing technology, many people around the world will need to re-skill to remain employable.

Here in the United States, post-secondary education is ever more out of reach for many as a result of increasing costs and competing priorities (like earning enough to keep pace with inflation). Low unemployment rates tempt people to enter the job market when they might otherwise pursue education. There's certainly nothing wrong with that, but in the long run, those who forego education may be sacrificing future earnings and opportunities.

Of those students who do start a college education, approximately 17 percent don't return for a second year.[30] Again, that number is not uniform across interest groups. Black and Native American populations leave at higher rates than other groups, at 24.7 percent and 37.3 percent respectively.[31]

These are just some of the challenges facing learners. On the other side of the coin, educational institutions are challenged to recruit, retain, support, and graduate students in an increasingly competitive environment. The pace of change is so fast that an incoming first-year undergraduate student may face a very different employment landscape four years later. How can students and institutions prepare under such circumstances?

AI and ML have some answers.

SUPPORTING MORE STUDENTS MORE EFFECTIVELY

In most learning environments, including schools, colleges, and universities, students easily outnumber teachers, advisors, and other staff. This presents numerous challenges for the personnel tasked with serving, supporting, and guiding the entire population of students at any institution, and the problems become more acute with scale.

Large public universities often enroll tens of thousands of students each year. Those students always have lots of questions, especially when they are new to a campus. The *Chronicle of Higher Education* shared an example from Georgia State University (52,000 students): "Timothy M. Renick, vice president for enrollment management and student success, notes that the financial-aid office receives as many as 2,000 calls a day from students in the weeks before start of each semester. 'We're not American Express,' he says. 'We don't have a call center with 200 people in it.'"[32]

Although some of those questions are probably complicated, unique, or nuanced enough to require a human response, many more are routine, even mundane, and may not need responses. Answering routine questions repeatedly is an inefficient use of a human advisor's time and knowledge, but that's little comfort to the students who need the information and struggle to find it elsewhere. This is a problem AI and ML can address. In fact, the

Chronicle of Higher Education reported, "Georgia State became the first university to work with AdmitHub, a company that has developed chatbots to communicate with college students through texting."[33]

Such a chatbot runs on software that predicts likely answers to questions users enter. If the software determines an answer—based on a sufficient volume of data and historical examples—and predicts that the answer is likely correct, the question is answered automatically. A confidence threshold can be set to determine just how likely the prediction needs to be. Georgia State University's system used a 95 percent confidence threshold; if the software's answer can't meet that, the question is redirected to a human.[34]

As you can see, a system like this alleviates friction on both ends of the exchange. The students asking questions get immediate answers, directly in the moment. Meanwhile, the personnel responding to questions are freed up to spend their time on more valuable tasks.

Georgia State University did not stop there, however; the school first integrated AI chatbot support directly into course content in 2021 and saw encouraging results. Receiving direct text messages about their class assignments, academic supports, and course content increased the likelihood students would earn a B grade or higher and, for first-generation students, increased their likelihood of passing the class. First-generation students receiving the messages earned final grades about eleven points higher than their peers.[35]

For perspective, consider that students at Georgia State University already received email reminders about courses, due dates, upcoming exams, and so on. With the chatbot, students "were getting an extra, direct reminder as well as the opportunity to text questions back . . . to their professor."[36]

Similar applications of AI assistants have been piloted elsewhere. In 2016, computer science professor Ashok K. Goel used IBM's Watson technology as a teaching assistant for an online course offered at the Georgia Institute of Technology. He named the bot Jill Watson and deployed it alongside nine human teaching assistants to answer questions from the 300 enrolled students. According to an article in the *Chronicle of Higher Education*, only a few students suspected that Jill was a computer program.[37]

The same article described how Goel trained the bot on a database of 40,000 questions from previous courses, which enabled it to respond to routine inquiries. In a course with hundreds of students enrolled, the professors and teaching assistants can receive thousands of questions during a single semester. If an AI-powered bot like Jill can answer students' routine inquiries about assignments, due dates, and basic concepts, that frees up time for professors like the authors of this book to innovate on content and pedagogy and ensures that students aren't waiting around for information they need in order to progress.

Newer advances in Generative AI will allow every professor and teacher to upload their content (slides, video recording, transcripts of lectures) and create their own personalized LLM-based teaching assistants for their courses. The Shorty (2016) and Webby (2012) award-winning California-based Khan Academy wan an early adopter of this approach. Their *Khanmigo* personalized tutor is designed to be a learning companion for students who may be struggling to solve a calculus problem or to engage with a book assigned in their literature class.[38] Like a great tutor, it won't give a struggling student the direct answer, to say, a calculus problem; rather it will nudge the student to learn the sequence of relevant concepts on the way to finally solving the problem on their own accord. Innovations such as Khanmigo and the one Goel designed for his course are likely to play an increasing role in scaling high-quality educational access to remote corners of the globe. For years, colleges and universities have packed lecture halls with hundreds of students enrolled in introductory courses. The recent trend toward massive open online courses (MOOCs) is another effort to try and scale education to meet the needs of thousands of learners. In both scenarios, teachers simply can't respond to all student questions without help. AI can provide that help efficiently and effectively.

AUTOMATED TUTORING AND SELF-PACED INSTRUCTION

AI can provide tutoring guidance to students as well, thus supporting their learning directly.

One researcher in this area is Vincent Aleven at Carnegie Mellon's Human-Computer Interaction Institute, whose projects include a study of

AI-powered tutoring for middle school students. His team of researchers "wants to understand if data-optimized AI-based tutoring software, smartphones and social motivation can work together to enhance learning."[39]

As that quote indicates, there's often more to designing a successful AI application than the algorithm alone. In this case, the researchers' interest in smartphones and the particular social motivations of their population (teens and preteens) are equally important.

By 2030, half of the world's youth will live in countries with mobile-first or mobile-only Internet connections.[40] Smartphones and mobile devices have the potential to reduce the cost of educational equipment, alleviate pressures on Internet access, and expand access to educational opportunities in ways that are affordable and accessible.

Moreover, several experimental studies point to a strong relationship between mobile technologies and learning. For example, Aker and colleagues found that adults in Niger who received weekly phone calls as part of a two-year education program improved their math and literacy test scores over those who did not receive weekly calls.[41] York and Loeb showed that in the United States, low-income parents who received three weekly text messages about their children's academic skills increased their own involvement in their child's learning.[42] These texts also helped to increase their children's gains in early literacy. Angrist and colleagues conducted a field experiment in Botswana to examine the effectiveness of mobile intervention for students' learning during the early lockdown period of the COVID-19 pandemic.[43] Students were randomly assigned to three groups: one group received SMS texts with weekly numerical problems and fifteen- to twenty-minute live phone-call walk-throughs of the problems. The second group received only the SMS texts, and the third group did not receive anything. They found that the group with SMS texts plus phone calls experienced a 24 percent gain on performance scores, whereas the SMS-only group experienced a 13 percent level gain on performance scores. Deng and colleagues recently designed a field experiment in a Chinese vocational school entailing random assignment of students into three groups: the first group was allowed no smartphone use in class, the second was allowed unlimited smartphone use in class, and the third was allowed unlimited smartphone use with teachers' instructional assistance.[44] Via a video feed, the authors measured performance scores before and after lectures as well as time spent in learning and time wasted in being distracted. They found that students were equally distracted across all three groups, but the teacher-assisted group spent a larger portion of the available time in learning.

To date, the research on the relationship between mobile technologies and learner outcomes is still thin. Whereas smartphones can bring tremendous potential to educational settings, particularly for disadvantaged groups crossing the digital divide, they also introduce opportunities for misuse and situations inducive of distraction, both of which, of course, degrade learning outcomes.[45]

One large, randomized control trial study published in 2016 found mixed results with respect to AI-based tutors. That study examined the use of "innovative automated tutoring software that provides self-paced individualized instruction and attempts to bring students to mastery of a topic before progressing to more advanced topics."[46] The context was an algebra course for middle school and high school students. The researchers conducted the study in fifty-one school districts across seven states, with 147 participating schools. At the conclusion of the two-year study, the results showed positive effects on exam scores after the second year (approximately eight points better), but only for the high school students, not the middle schoolers.[47]

PERSONALIZED LEARNING JOURNEYS

Another advantage of data-driven AI is the opportunity to personalize content and experiences for each individual. Personalization is already ubiquitous in the technology we use to shop (think Amazon, Google, Meta, and other recommendation-driven engines), choose our entertainment (Netflix, Spotify, Pandora), decide where to go and what to do (location-based recommendations), and more. In the future, personalized learning will be just as common.

The country of Estonia is already heading there. Its government supports various projects designed "to implement AI-driven solutions to personalize students' learning paths."[48] Most traditional educational systems and institutions have grouped students mainly by age and provided roughly the same instruction to all students in a given group. Resource constraints—namely teachers' time and energy—kept personalization at a minimum in most cases.

In Estonia, the government "has started building a personalized learning path infrastructure" using data and AI.[49] The effort includes multiple projects incorporating AI and machine learning in diagnostic testing, customization of materials, and self-paced learning. As *Education Estonia* reports: "What is great for one student, may not be good for another. What improves the progress of an 'average student,' may hold back top performers."[50]

Estonia's approach is attracting attention and praise. The former Soviet state's education system was named the best in Europe by the Programme for International Student Assessment (PISA), an educational assessment and innovation effort created by the Organisation for Economic Co-operation and Development (OECD).[51]

Personalized, computer-assisted learning can further help students in places where qualified teachers aren't always available, such as in developing countries and more remote communities. A study of middle school students in India showed that students who received personalized homework achieved better exam scores (4.16 percent higher) compared to students who did not get personalized homework.[52] Combined with the spread of mobile technology and supportive policies (such as those adopted by Estonia), computer-assisted learning can be used to deliver education to entire populations efficiently and effectively.

In the India study, the effect was not uniform across ability levels: "The personalized homework benefited medium ability students but not the high and low ability students."[53] But varied findings like this are not really cause for concern. On the contrary, research that reveals these nuances makes it possible to design highly targeted (i.e., personalized) instruction and support.

PREDICTING STUDENT RETENTION

The twenty-first century has brought a new round of challenges to institutions of higher education. Increasing competition, shifting expectations, and rising costs threaten to diminish or even topple once venerable institutions.

Data often hold clues to solving the messiest problems that arise from these complex systems populated by thousands of people.

For example, colleges and universities want to see all the students they admit succeed and graduate. But for a wide variety of reasons, a (hopefully) small portion of enrolled students will drop out or leave any school each year and not return. This is a well-known occurrence, and administrators have expended enormous amounts of money implementing programs aimed at maximizing retention rates. Among campus populations in the thousands and tens of thousands, it's Very difficult to know *which* students are likely to drop out. As a result, retention efforts are often broadcast to the entire population, which is both a waste of resources and not always effective at reaching the students who need additional support.

Using the data colleges and universities have on their students, machine learning models can predict which students are most likely to drop out. This approach allows administrators to intervene *in advance* using advising and other support services, thus ensuring that the students who can actually benefit from the resources receive them. This is good for the institutions (retention rates impact rankings and reputations) and students alike. The power of machine learning in this scenario is that, given enough data, predictive models can spot patterns that humans cannot. Traditional indicators such as grades or surveys sent to students are limited in this regard, since grades don't tell the whole story and students may not respond to surveys, among other factors.

Professor Sudha Ram at the University of Arizona Eller College of Management showed the potential of a data-driven prediction model in 2017. Ram and her team used a different kind of data: the digital trail left when students swipe their ID cards at locations and services around campus. "Students swipe smart cards at such locations as the gym, library, bookstore, and food court, activities indicative of social integration. More than dropping grades, Ram says, student social interactions indicate whether first years are comfortable, and therefore the likelihood that they will remain or drop out."[54] The model successfully predicted nearly 90 percent of at-risk students.

No human analyst, or even a large team of analysts, could ever analyze that volume of data, let alone find the relevant patterns in time to make a difference.

THIS IS JUST THE BEGINNING

So far, we've explored some examples of how AI and ML can answer student questions, save teacher time, deliver support services, predict which students are at risk of dropping out, and even tutor individuals on their personal learning journeys. But this list is not exhaustive.

Applications of artificial intelligence in education are plentiful. AI helps schools reduce instances of cheating (especially online).[55] It helps institutions run more efficient facilities and campuses.[56] There's more: AI can enhance accessibility for learners with different needs, teach preschoolers basic academic skills through interactive games, and help administrators optimize schedules and lesson plans.[57]

AI brings much more to education than a bot that's good at writing passable (if mediocre) text filled with "accurately simulated natural-language sentences."[58] The reality of AI is much wider and more diverse, and it's evolving rapidly. Just as the personal computer changed much of our educational systems, AI is positioned to usher in a new normal that we are only beginning to glimpse.

With such fundamental changes come significant challenges as well. Teachers, professors, school leaders, and parents are going to have to adapt. We are going to have to shift how we teach and how we assess student achievement. Rote memorization of facts and written essay answers are going to mean even less than they do today—we've had the ability to look up facts on demand for decades; now we have ChatGPT to synthesize facts into workable structures. Perhaps we won't even rely on written tests as demonstrations of mastery. As Timothy Burke writes: "There's one thing you can't use an AI to do, and that's demonstrate live and in-person that you know your stuff to a human instructor who is a trained expert in what you've learned."[59]

"A.I. will probably give us fantastic tools that will help us outsource a lot of our current mental work," wrote *New York Times* columnist David Brooks. "At the same time, A.I. will force us humans to double down on those talents and skills that only humans possess."[60] That's progress.

TAKEAWAYS

- Powerful Generative AI–based LLMs such as GPT-4 have the ability to replicate human abilities of writing, exam-taking, drawing, and creating. GPT-4, like previous language models, is trained to predict the next word in a document using publicly available data (think of all the Internet content and then some). LLMs have the power to upend education as we know it and are a source of much concern for many educators. We believe that while there will be a period of adjustment, Generative AI will be harnessed to augment and enhance the learning potential of many.
- There is no shortage of grand challenges to solve, and increasing education levels globally is one of them. If LLMs can be harnessed—with the appropriate guardrails—to individualize tutoring (recall the Khanmigo example) or decrease writer's block, then AI can be a source of significant productivity for humankind. Just as calculators became an important tool for students in math classes, ChatGPT has potential to become an important asset for students who want to learn better, get tutoring, and hone their creative and critical thinking skills.
- Using AI-powered chatbots (recall the Georgia State University example) to communicate with college students can improve their experience by dealing with routine inquiries about assignments, due dates, and basic concepts. This can free up time for faculty to innovate on content and pedagogy and ensures that students aren't waiting around for information they need to progress. Similarly, AI can help predict which students are most likely to drop out (as in the University of Arizona example), allowing administrators to intervene in advance using advising and other support services, thus ensuring that the students who can actually benefit from the resources receive them.

6 WORK, CAREER, AND FULFILLMENT

You met our (fictional) friend Jasmina in chapter 1. Based in Phoenix, Jasmina has a thriving career in sales for a software company. She's worked hard to deliver results for her employer, and her efforts were recently rewarded with a promotion. When offered the chance to take the lead on sales in Texas in addition to Arizona, Jasmina welcomed the opportunity to continue building her reputation in the industry. She felt ready for a new challenge.

Jasmina tackled her new role with gusto. She made two or three trips to Texas each month to familiarize herself with the market, follow up on leads, and meet with prospects she hoped to turn into clients. She developed a fondness for *migas* at breakfast and landed a respectable slate of new customers as the months passed. She was proud of her success, and her employer was more than satisfied with her performance. But as her first anniversary in the role came and went, Jasmina started to question whether the demands of the job were beginning to outweigh the benefits.

First and foremost, the demanding travel schedule took Jasmina away from her son more than she had originally expected. Juggling his schedule and her own had always been a challenge, but this was a new level of complexity as she struggled to remain available for everyday life priorities, such as watching her son compete at his swim meets, taking weekend hikes with her sister, and trying to sustain a romantic relationship.

These concerns had been simmering in Jasmina's mind for a few months when the COVID-19 outbreak hit the United States in March 2020. She had just returned to Phoenix when the shutdowns started. Like so many others, her company closed its offices and sent everyone home. Her son's school did the same, and within days it became clear that Jasmina would not be traveling anywhere for a while. Her upcoming trips were all cancelled. Moving forward, all sales activities would be conducted remotely via email, teleconferencing, and video meetings.

Jasmina was still adjusting to this new reality when rumors of impending layoffs began to circulate among her colleagues. She braced herself for this possibility, and even when she survived the first round of cuts, she knew her job was no longer secure. Her stellar past performance reviews would mean little if the business didn't survive—a very real possibility in the face of a global pandemic.

Sequestered at home, Jasmina divided her time between Zoom calls and helping her son navigate his e-learning lessons. She watched news coverage of overtaxed hospital workers and exhausted cleaning crews spraying disinfectants. She ordered masks online. Weeks turned into months. She thought a lot about her son's future as well as her own. What would life look like on the other side of these historic events?

When Jasmina heard the news that her company had been acquired by another firm, she wasn't surprised. So much had changed during the pandemic; was it any wonder that the company's leadership team would accept a buyout offer given the uncertain and challenging circumstances?

The acquiring company's VP of sales offered Jasmina a position in the newly merged organization, but the thought of adapting to a new company culture—one she hadn't chosen for herself—left Jasmina cold, disheartened even. She didn't necessarily want to hit the road again; her experience in Texas had exhausted her appetite for work-related travel. Even more than that, she realized she abhorred the idea that her days would continue to be filled with tedious video calls and endless email exchanges about contract minutiae. Outgoing by nature, she craved face-to-face interaction with other people and the feeling she got when she helped someone solve a problem. Maybe she could find more of what she wanted in another job, perhaps even another field entirely. She declined the offer with the new organization and took the severance package rather than continue in a job that would drain her.

When Jasmina reflected on everything she'd been through leading up to and during the pandemic, she concluded that she wouldn't just search for another sales job. She resolved to find a whole new career. A lingering concern for her was how her career transition would fare in a job market powered by AI and algorithms. She was aware of the scandal caused by Amazon's AI-based recruiting process that favored men over women for tech jobs (mentioned in chapter 1). Being a person of color, would she fall victim to AI-based resume screening likely to (a) assume that groups with proven track records from the past are (solely) likely to perform well in the future, and (b) be biased because certain resume writing style and socio-linguistics correlate with protected characteristics such as ethnicity, race, and gender.[1] Would the algorithms currently in use be able to go beyond traditional approaches taken by recruiters and potentially increase her chances of faring well in the job market? And would recent advances in Generative AI–based conversational assistants change the nature of work for certain types of workers more than for others?

WORK: AN EVOLVING LANDSCAPE

How the modern world works has changed dramatically in recent years. Certainly, the COVID-19 pandemic was an external shock to the system, one that disrupted entire industries and decades of established employment practices. You may have experienced some of this firsthand. Perhaps you lost a job during the pandemic, witnessed your company shift to remote work, or continued to work under hyperstressful conditions if your job was deemed essential.

So much changed so quickly that it will be a while before society fully understands the pandemic's long-term impacts on work, employment, and economics, but most observers agree the pandemic accelerated existing trends as well as initiating new ones and halting others.

That isn't to say that COVID-19 is solely responsible for all the changes in these areas. On the contrary, when it comes to the evolution of how we work, many other forces are at play: consider globalization, automation, economic inequality, the gig economy, the expanding role of knowledge work, weakened labor unions, lower birthrates and other population trends, technological innovations, and shifting trade alliances, to name but a sampling. On top of this long list, we can add the recent AI developments in large language models that have given us human-like conversational agents such as GPT-4 from companies such as Open AI. The employment landscape and the rules of engagement are often in flux, rarely fixed. These developments and others predate the pandemic disruptions of 2020 and 2021 and they haven't abated.

Sometimes the pace of all this change can feel dizzying for those affected, especially when monumental shifts happen within a generation, or a decade, or an even shorter time span. After all, most people need to work throughout their adult lives, and many work past traditional retirement age by choice or necessity. Whether you recently entered the workforce or are established in a career, the working world you'll navigate looks very different from the one your parents or grandparents inhabited.

Of course, change is not inherently bad; change is a necessary ingredient for progress, advancement, and evolution. And according to some, the working world was long overdue for significant change.

For example, human resources veteran and *Forbes* contributor Liz Ryan wrote in 2018, "The hiring process has been breaking down for twenty years, but the pace of its devolution has been picking up lately." Responding to a letter from a frustrated job hunter, Ryan lamented, "More and more employers are adopting pointless, expensive and time-consuming extra processes to make their recruiting pipelines even slower and more off-putting to candidates." In particular, she identified "endless pre-employment tests and questionnaires, keyword-searching algorithms used to screen resumes instead of human eyes and human judgment, automated one-way interviewing systems and a host of other 'interventions'" as culprits for the breakdown.[2]

Time-consuming, onerous hiring processes aren't the only problem. According to another 2018 *Forbes* article, "About 65 percent of people look for new jobs within 91 days of being hired."[3] That statistic was supplied by Raj Mukherjee, a senior vice president at Indeed.com, the large employment matching platform. So many discontented employees suggest major issues related to job satisfaction and employee retention. And that article was published years before the Great Resignation of 2021.

Like the story of Jasmina resigning rather than continue in a job she no longer enjoyed, millions of workers quit their jobs in 2021. This fact alone doesn't capture the magnitude of the resignations; the numbers are historic. According to the U.S. Bureau of Labor Statistics and the Society for Human Resource Management, 47.8 million workers quit in 2021, "the largest exodus of employees on record."[4] That's an average of about four million people quitting each month. The next closest year on record is 2019 with an average of about 3.5 million resignations per month. In addition to these averages, there's a notable difference in the trends for these two years. The line for 2019 in the article's interactive chart showing monthly resignations year by year is nearly flat. In other words, the number of people quitting each month was roughly the same across the year. But in 2022, the line slopes noticeably upward, suggesting a growing trend.

Change can often be stressful, especially when it comes to one's livelihood. Our friend Jasmina likely entertained the idea of looking for a new job for some time before making the decision to resign. Leaving the known for the unknown always involves risks, and a savvy professional like Jasmina (or you) knows that job hunting can be challenging. Like many others in her situation, Jasmina had worries: In today's marketplace, where can one find the best opportunities? Do hiring managers even *look* at resumes anymore? Many big companies require application via a computer portal that feeds into AI algorithms; if all the resume screening is being done by algorithms, what can an individual do to increase their chances of landing an interview with a real person who can make a decision? Do those changing their careers like Jasmina have any hope of transferring their skills to a new profession? Or, as mentioned earlier, will the newest AI screening algorithms disregard a woman of color like Jasmina out of some hard-coded bias?

All of these questions and fears are legitimate. The popular media loves to stoke our anxieties when it comes to losing control, especially to a cold, soulless computer. But in the end, what matters are the *answers* to these questions, and you may be surprised to learn that AI, algorithms, and technology platforms bring a variety of opportunities and benefits that are already shaping various aspects of work. There are several positive developments to celebrate as a result. For instance, a recent National Bureau of

Economics Research working paper examined the impact of a staggered introduction of a Generative AI–based chatbot to augment the abilities of 5,179 customer service agents.[5] Access to the tool increased productivity, as measured by issues resolved per hour, by 14 percent on average, with the greatest impact on novice and low-skilled workers, and minimal impact on experienced and highly skilled workers. The underlying mechanisms leading to this increase in productivity were found to be an improvement in customer sentiment, reduction in requests for managerial intervention, and improved employee retention. Armed with this insight from current research let us take a step back and look at some of the ways AI is *improving* how employers recruit and hire employees and how people find career opportunities and fulfillment.

HELP WANTED: APPLY WITHIN

If you take a moment to reflect on your own job-hunting experiences, what comes to mind? Did you come of age during a time when businesses displayed "Help Wanted" signs in windows or listed notices in newspaper classified ads? Perhaps you remember hours spent scouring online job boards, reading through position descriptions and lists of requirements until your vision blurred. Maybe you worked with a professional recruiter at some point, hoping that person had some magical insight into a field you wanted to enter. Are you a diehard believer in the "it's who you know" school of thought, and therefore put all your effort into networking in hopes of meeting the person who could offer you a dream job? Or maybe you're an introvert who hates gladhanding strangers at industry events on the off chance that it leads to an opportunity, so you stuck to just asking friends and family if they knew of anyone who was hiring.

It's easy to see what all of these examples have in common: they all rely on an element of randomness that is both inefficient and unlikely to yield optimal results. A great job is out there *if* you happen to walk by the sign in the window or open the newspaper on the right day. You'd apply for that open position *if* you hadn't missed the deadline because you were too busy to weed through hundreds of online postings that week. You could have landed an interview for a dream job *if only* you'd sat down at a different table during that luncheon you attended.

That's a whole lot of "ifs" messing with your chances.

AI offers alternatives to the random nature of traditional job hunting. Integrating AI into recruiting and hiring benefits both job seekers and employers in various ways.

FINDING THE RIGHT FIT

AI and machine learning are well suited to tasks that require searching through dense collections of data points for meaningful patterns. This ability can be used to match people with more suitable options, whether in romantic partners (as we saw in chapter 2 on dating) or jobs. We're not the only ones to make this comparison. "To put it simply, AI can help you speed date open positions to find your next job," says Gergo Vari in *Fast Company*.[6] He continues:

> AI is alive and well in the realm of dating. People who sign up on sites like Tinder don't just educate the platform about the kind of person they're looking to connect with. They also share more granular insights through their search patterns, their connection requests to others on the site, and the messages they respond to (or ignore).
>
> It doesn't take long for the dating platform to become proactive and accurate when it comes to sending you suggestions about who to connect with. It already knows you may be interested in those individuals. Ultimately, it's up to you to act responsibly and decide whether you'll play the hand AI deals you, but it takes away much of the initial guesswork. The same concept applies to job search when AI is used effectively.

For job seekers, the benefits of AI are many. Instead of manually researching companies and reviewing job requirements, customizing resumes, and wasting time applying for jobs that ultimately turn out to be a poor fit, AI-powered hunting can match a candidate's skills to appropriate positions. Per *Forbes*, "As AI and machine learning develop in the field, a service like Indeed should be able to suggest new, much more compatible opportunities based on a job seekers' [*sic*] work experience, skills, salary, interests and location."[7]

As a result, good opportunities are surfaced, and poor ones are filtered out. Applicants can spend their time pursuing the *right* openings (i.e., ones more likely to yield an offer) instead of dead-ends.

This greater efficiency becomes even more meaningful when we consider that the number of potential jobs a person might qualify for has increased significantly for many people. The pandemic forced a rapid, and largely successful, shift to remote and asynchronous work for a variety of jobs. Companies around the world learned that physical presence in an office was not necessary for many types of work, so the rules and norms governing corporate cultures finally caught up to where the technology had long been. Remote work became the norm rather than the exception.

Having experienced this model during the pandemic, many employees who *can* work remotely don't want to return to stressful commutes, rigid schedules, or cubicle farms. As the pandemic wanes, leaders at all kinds of companies have realized they will need to continue offering the flexibility of remote work where possible. Otherwise, they risk losing workers to companies that do. Researchers surveyed thirty thousand Americans and concluded that some form of work-from-home is here to stay for a variety of reasons, including increased benefits for employees, especially those with higher earnings.[8] Thanks to this greater acceptance of remote work, job seekers are less limited by geography and transportation. Instead, they can consider jobs in faraway places and may never worry about contacting a moving company.

Similar AI tools can be used to help employees find a desirable work *culture*, in addition to the job itself. Finding a company where the values and working style match one's own can have a big impact on job satisfaction and how long someone stays. Others have observed, "To ensure candidates remain engaged and committed to the jobs they're offered, talent firms are leaning on AI to match job hunters' organizational-culture preferences to companies that align with them."[9]

Candidates in today's market are taking note of companies such as Unilever that processes millions of applications while recruiting more than thirty thousand people a year. "As a multinational brand operating in 190 countries . . . Unilever can't afford to overlook talent just because it is buried at the bottom of a pile of CVs," writes futurist and author Bernard

Marr.[10] At that scale, automation is essential, and AI-powered systems like the one Unilever uses can speed up the process and save significant time and money. Leena Nair, Unilever's head of HR (human resources), tells Marr that automation cut 70,000 person-hours from the recruiting process, freeing up time for their recruiters to focus on building relationships with candidates, which should benefit both the company and the candidates.[11]

Casting a wider net for talent is quickly becoming an imperative for many businesses. At the time of this writing (early 2023), the environment for employers looking to hire is quite competitive across industries. The Great Resignation created millions of job vacancies, unemployment is historically low, and candidates have more options. Maxime Legardez Coquin, CEO of a company that makes skills-assessment software, puts it this way: "The Great Resignation and current job market have shown that people are realizing how much control they have in the workforce and are feeling empowered to not immediately jump after every employer that comes their way."[12]

In addition to screening more candidates, AI-powered hiring can help companies find the best people for the jobs that need to be filled. This is the same two-way street that helps individuals find attractive job opportunities. By analyzing troves of data in applications, personnel records, and more, AI and machine learning can find relevant patterns that indicate what kinds of candidates are well suited to what kinds of jobs. Even incremental improvements can save time and money for candidates and employers alike and increase the chances of successful employment decisions.

Another kind of efficiency produced by AI was demonstrated by a team of business school academics.[13] The authors developed an AI model for salesforce hiring using recordings of conversational video interviews that involve two-sided, back-and-forth interactions with messages conveyed through multiple modalities (text, voice, and body language). This is an excellent example where AI tools are used on conversational videos to capture features related to two-way conversational interactivity and human body language. This not only helps increase efficiency in hiring, but also helps expand the geographic pool from which people can be hired. A company in New York can essentially hire someone based on a recorded video interview from dozens of cities in the United States.. Going forward such video analytics-based applications will significantly help candidates in the labor market by expanding the pool of recruiters.

THE TRANSITION TO SKILLS-BASED HIRING

A major reason for AI's impact on recruiting is that it is not limited by traditional tools like the resume or CV. Many employees and employers alike view such devices as outdated and inadequate, and for good reason. According to Coquin, "For too long, the hiring process has been based on whether someone has previous experiences at the right company or whether they went to the right school to get the right degree and not the most important question at hand: Does this person have the right skills to actually do the job?"[14]

Skills assessment and matching via machine learning are well on the way to yielding benefits for all involved. For example, "by focusing on skills, recruiters open the door to applicants with atypical experiences (e.g., trade schools, military service, online courses, volunteering, etc.)," says Coquin.[15] A study published in the *Harvard Business Review* looked at fifty-one million jobs posted between 2017 and 2020 and found that employers are deemphasizing degrees in favor of skills, especially for IT and managerial positions.[16] In other words, companies can tap into new sources of talent to meet their hiring needs, and at the same time, people who might have missed out on opportunities under the traditional model can break into fields that interest them and where they are likely to succeed.

The same techniques can help people transition from one field or occupation to another, a major plus in the ever-changing modern work environment. This benefits individuals who choose to shift careers, like Jasmina. And there are many people who fall into that category, as the Great Resignation attests. Over the course of their working years, millions of workers may find themselves forced to change careers, fields, industries, or all these. If automation makes your job obsolete or economic shifts cause a downturn in your industry, you'll want whatever transition you make to be as quick and smooth as possible. AI is ready to help.

One study published in 2021 illustrates AI's ability to identify which skills are best suited for various occupations.[17] The authors compiled data gleaned from job ads in Australia as well as employment statistics from the Australian Bureau of Statistics (ABS) for the study. Using this data, their first step was to "measure the distance between sets of skills from more than 8 million real-time job advertisements (ads) in Australia from 2012–2020." The idea is that "when two skill sets are highly similar (for example, two occupations), the skills gap is narrow, and the barriers to transitioning from one to the other are low." In other words, if the skills needed to perform job A are *highly* similar to the skills needed for job B, but not so similar to the skills needed for job C, a worker who possesses job A-type skills will more easily transition to job B rather than C.

The logic here is simple enough: Max knows how to fix automobiles and Carlos has experience as a restaurant inspector. Based on skillset similarity, Max's talents are more transferable to a job as a maintenance technician for factory equipment, whereas Carlos could successfully assume a role as a compliance officer in another industry. But this simple logic is just the starting point.

What's interesting and powerful is the fact that AI and machine learning can find similarities that are not obvious to humans. Spotting the skills match in the deliberately simple example of Max and Carlos is easy, but determining which skillsets are a closer match to those needed for a job coding new software, designing a municipal water filtration system, or managing a chemistry lab is not simple at all. Here our human intuition and experience are little more than random guesses.

In the Australian study, machine learning looks at all the data points, makes millions of calculations, and determines similarities at a speed and scale unachievable by even the most seasoned human resources managers. The resulting similarity scores are not limited to two or even three dimensions; rather, the scores consider multiple dimensions, far more than can be easily held in the human mind. Using the relative similarity scores, the researchers then use more machine learning to build a recommender system, one that tells us which job transitions—among millions of potential options—are more or less likely to lead to success.

In the end, data-driven and AI-powered methods like those demonstrated in the Australian study can positively impact the economic prospects of millions of workers, and by extension, the communities and markets in which they live. People like Max, Carlos, Jasmina, really *all of us* will have an easier time returning to the workforce after an event like COVID-19 forces us out or when other circumstances in our lives prompt us to pursue a different career path by choice.

LESS BIAS, MORE FAIRNESS

Another area where AI is improving our working lives centers on an important social issue, which was top of mind for Jasmina as she weighed her career transition plans. Bias in hiring has long denied individuals from certain races, ethnicities, and genders equal employment opportunities. This has contributed to severe economic consequences including disproportionate unemployment rates, loss of generational wealth, unequal pay for women, and much more. In places like the United States, where healthcare is tied to employment, bias in hiring has far-reaching ripple effects. No job, no (or expensive) healthcare.

A relatively recent study explores some of the mechanisms and implications of bias in recruiting.[18] The study's authors cite several sources showing that "resumes containing minority racial cues, such as a distinctively African American or Asian name, lead to 30–50 percent fewer callbacks from employers than do otherwise equivalent resumes without such cues," and conduct further research on how job applicants adapt to such a labor market.

In their work, the researchers examine both the means of "whitening" (e.g., changing names, omitting certain kinds of experiences) as well as the motivations for or against doing so (e.g., to "get a foot in the door," belief in meritocracy, or as a tactic to avoid discriminatory work environments). The study concludes that for both African American and Asian applicants, "There is a clear pattern across both groups: whitened resumes led to more callbacks [from employers] than unwhitened resumes." Other studies have found that "[white-sounding] names receive 50 percent more callbacks for interviews" than African American-sounding names, a racial gap that is "uniform across occupation, industry, and employer size."[19]

Discrimination in hiring is not limited to applicant race. Research has shown bias against women,[20] LGBTQ+ individuals, and older applicants[21] as well as stigmatized groups such as ex-offenders.[22] Much of the bias is in fact unconscious—well-meaning recruiters and HR managers don't intend to discriminate, but biases emerge, nonetheless.

AI can mitigate and even eliminate bias in hiring. As Frida Polli writes in *Harvard Business Review*, AI's ability to review many more potential candidates is part of the solution—AI doesn't have to rely on arbitrary cutoff mechanisms like human recruiters often do.[23] In addition, Polli argues, "A beauty of AI is that we can design it to meet certain beneficial specifications," meaning that AI systems can be trained to counteract biases that are inherent to human decision-making.

A particularly promising approach is the use of reinforcement learning to design resume-screening algorithms. We briefly introduced this in chapter 1, but it is worth diving a bit deeper here to build your intuition into its inner workings. Imagine you are starting a chain of neighborhood hardware stores in predominantly white neighborhoods in the Upper Midwest (where one of us lives). You have had great success in your initial year hiring help from the local high school, which, no surprise is also predominantly white. Any time a new opening is posted, say for a B2B business development manager, and you are screening resumes you have a decision-making choice. You can take the "best" action from your current worldview—hire applicants who look like those who have worked well for you in your initial year (we call this an *exploit* strategy)—or you can choose to deliberately try to learn about other actions (we call this an *explore* strategy) that *might* be better than your current best strategy. The key though is that you don't want to randomly explore—after all, you have customers to please. You want to explore those candidates, say from a neighborhood across the highway, which is predominantly Black, who have high potential but also high uncertainty. Reinforcement learning algorithms such as the Upper Confidence Bound are designed to help you exploit and explore at the same time and are showing great promise in increasing the diversity that results from resume screening.[24]

We predict that a particularly impactful disruption in work and career paths is going to be caused by Generative AI technologies. A handful of academics have already begun to quantify the effects of Generative AI on the labor market. Felten and colleagues have predicted which industries will be most affected by LLMs in the future by using a measure of compatibility between the required tasks in an industry and AI's capabilities over the 2010 to 2015 time period.[25]

A recent set of papers conducts field experiments randomizing access to Generative AI tools and studies the effect on worker performance. For example, in 2023 Noy and Zhang tested how well ChatGPT helps people with professional writing tasks.[26] They gave 444 college-educated people specific writing jobs and let half of them use ChatGPT. The results? Those using ChatGPT wrote faster and produced better work. The chatbot was especially helpful for less skilled workers, making the gap between top and average performers smaller. Instead of making workers work harder, ChatGPT allowed them to focus on coming up with ideas and editing, rather than just writing a first draft. People who used ChatGPT enjoyed their work more and felt more capable, but they also had mixed feelings about machines doing their jobs. Another set of researchers led by Yilmaz showed that the introduction of Google's machine translation reduces translators' employment, especially for tasks with analytical elements.[27]

AI-generated art is dramatically changing the art business and the occupation itself. Anindya Ghose and his coauthors Hongxian Huang and Runshan Fu studied how Generative AI affects artists on two major online art sites—Lofter and Graffiti Kingdom—based in Asia.[28] On the one hand, they found that when Lofter started using Generative AI technologies for art, fewer artists used the platform. On the other hand, when Graffiti Kingdom stopped using Generative AI technologies, more artists returned to the site. This work demonstrated that the adoption of Generative AI art decreases artist activity on these platforms.

Academics have also begun to examine if LLM tools like ChatGPT can replace market research. If the model learns from customer reviews and blog posts, maybe it can also guess what consumers think. This could prove to be an effective way to do market research. At a recent workshop hosted by academics from the Wharton School of Business, several papers looked into this. These papers often discussed factors such as a person's income or details about a product, such as laptop specs and cost. They asked ChatGPT market research-like questions, such as "Will you buy this laptop?" They then checked whether ChatGPT's answers matched real people's answers. In short, ChatGPT results for market research are inconsistent. Yet, there are hopeful signs. Fine-tuning AI with the right data could lead to affordable, quick market research. It won't replace detailed studies but may be useful when time and money matter, even if it's not perfectly accurate. For a good summary see the September 15, 2023, blog post on LinkedIn by Kartik Hosanagar, one of the organizers of the workshop.[29]

The early evidence from the aforementioned ongoing research points to the importance of a new skill known as *prompt engineering*. This skill is about crafting good questions, providing rich contextual information, and relaying a well-thought-out sequence of instructions that communicate your intent to a Generative AI model. It will be the key to getting accurate, relevant, and coherent responses from Generative AI.

OVERCOMING OUR FEARS AND GETTING COMFORTABLE WITH CHANGE

If the idea that computers will influence or determine whether you land your next job interview makes you nervous, you're not alone. Pop culture is saturated with stories of humans falling victim to soulless machines usurping decisions people used to make. One of the best-known examples can be found in Stanley Kubrick's 1968 film *2001: A Space Odyssey* (based on an Arthur C. Clarke story). As the film progresses, the HAL 9000 computer that is supposed to *help* the crew gradually becomes something sinister and malevolent. The same theme appears in many sci-fi films and TV shows, from *Star Trek* to *Doctor Who* to *The Terminator*.[30]

Such fears of technology and computers have echoed through the media's coverage of today's AI. Sometimes the skepticism is thoughtfully expressed. For example, the AI gettingInstitute claims, "The vast majority of AI systems and related technologies are being put in place with minimal oversight, few accountability mechanisms, and little information about their broader implications."[31] A *Washington Post* story objects to specific companies like HireVue, makers of AI-powered recruitment platforms, criticizing such products as "unfair," "deceptive," "disturbing," and "digital snake oil."[32]

However, a lot of the chatter often boils down to the idea that AI is "just creepy."

"Few things seem creepier than algorithms mining our voices or photos to determine whether we should be considered for a job, and yet we're not that far from this scenario at all," wrote Tomas Chamorro-Premuzic and Reece Akhtar in 2019.[33] But they followed that sentence with, "One of the major problems with the way we currently interview job candidates is that the process is largely unstructured, leaving the questioning to the whims and fancies of the interviewer." This is not only inefficient, but also a good reminder that the status quo is not some egalitarian utopia.

In one instance of media coverage, *ProPublica* caused a stir with the headline, "There's software used across the country to predict future criminals. And it's biased against blacks."[34] The story criticized the use of AI in the criminal justice system. The journalists may have had the right intentions and wanted to shed light on prejudice and injustice. We have no argument with that. They rightly point out that the algorithm's scores skewed toward lower-risk categories for white defendants compared to Black or African American defendants.

Such biases *should* be pointed out, wherever they exist, so that they can be addressed and corrected where possible. As mentioned in chapter 1, Amazon discontinued its AI-based resume-screening process once it was revealed to reinforce bias against women.[35] Machine learning algorithms allow for this, when designed using ethical guidelines, but blaming the AI tool misses the point.

"This fear of biased AI ignores a critical fact: The deepest-rooted source of bias in AI is the human behavior it is simulating," as Polli puts it. "If you don't like what the AI is doing, you definitely won't like what humans are doing because AI is purely learning from humans."[36]

Chamorro-Premuzic and Akhtar concur. "AI algorithms simply leverage the same cues that humans do. The difference between humans and AI is that the latter can scale, can be automated, and if programmed correctly, will treat each candidate equally."[37]

And Loren Larsen, former chief technology officer at HireVue has argued that its AI system "is still more objective than the flawed metrics used by human recruiters."[38] Again, the research shows that unconscious bias *definitely* affects human recruiting decisions. But with AI, it's actually easier to identify and fix such bias.

Scholars have long been concerned about algorithmic bias and have pointed to several troubling examples in which algorithms trained using historical data appear to codify and amplify historical bias. The solution to this problem is artificial intelligence itself, guided by humans. Ethicists, regulators, and legislators have already proposed means and methods for using AI responsibly; for example, by requiring that AI systems be transparent and audited for bias.[39]

Academic research has already shown that bias can be mitigated while simultaneously delivering superior results. A field experiment by Bo Cowgill showed that machine learning technology could pick job candidates who were more likely to pass an interview, accept a job if offered, and be more productive once hired.[40] The key nuance is that the algorithms remove human biases exhibited in historical training data, when the human training decisions are sufficiently noisy. This occurs for nontraditional candidates who graduated from non-elite colleges, who lack job referrals, who lack prior experience, whose credentials are atypical and who have strong soft skills. This is exactly the profile that suffered under human decision-making but benefits under AI.

AI CAN BOOST JOB SATISFACTION AND CAREER FULFILLMENT

We've spent much of this chapter on the topics of job hunting (for individuals) and recruiting and hiring (for employers). But AI can help improve our working lives in several other ways.

A report from the IBM Smarter Workforce Institute summarizes how AI can help employees feel more *motivated* through engagement analysis, more *appreciated* thanks to smarter compensation planning, and more *valued* through customized growth and development opportunities.[41]

Ultimately, the combination of AI-powered recruiting, hiring, onboarding, and development can have a cumulative effect for individuals. More opportunities, better matches, easier transitions, and proactive help growing one's career are all about satisfaction and fulfillment. After all, work is not just about making money; work is also about self-respect, independence, and the rewards that come from contributing productively to the world and having a purpose.[42]

This may sound a bit romantic or Pollyannaish, but there are good reasons to reorient our relationship with work. For starters, there's our health. Although most modern workers in developed economies have escaped the unsafe and unhealthy conditions of yesteryear's factories, slaughterhouses, shipyards, and so on, there is evidence that many of our accepted work practices are still making us sick. Stanford professor Jeffrey Pfeffer explores this topic in depth in his book *Dying for a Paycheck: How Modern Management Harms Employee Health and Company Performance—and What We Can Do About It.*[43]

Of course, people do care about money—how they earn it and how much they get for the time and effort they spend. Perhaps then we should end with a quote from another bestselling author, Tom Rath:

> Even when money and your finances are an acute priority, it literally pays to focus on the value you're bringing to others. When researchers followed a longitudinal sample of 4,660 people over nine years, they found that having a sense of purpose in the first year of the study (based on a standard assessment of purpose in life) was associated with higher levels of both income and net worth over time. What's more, even when they controlled for other variables like life satisfaction and socioeconomic status, people with a sense of purpose at work also had significantly higher incomes at the end of those nine years.[44]

The good news is that AI can help all of us make progress on both fronts, paycheck and purpose, simultaneously. Gen Z are much more focused on work-life balance and self-care than past generations. And the advent and ubiquity of AI are helping them achieve their goals.

Our aim with this chapter was to demonstrate that questions and skepticism about the role AI and machine learning play in hiring decisions are natural, legitimate, and important, but once the questions have been asked, what matters are the answers. Current research and evidence suggest that there is more upside and less downside than one might assume. In fact, given what we know today, it's very possible that AI and ML will help you find your next job, a more fulfilling career, a guiding purpose, or all of the above. Along the way, AI and ML show great potential to help reduce bias and discrimination in hiring while going beyond the capabilities of human recruiters to match skills to jobs. In a few years' time, this could result in many more happy workers and a more favorable economic environment for us all.

TAKEAWAYS

- The employment landscape is laden with a variety of frictions and is often in flux. The advent of Generative AI is the latest addition to the long list of forces upending the world of work. An estimated 65 percent of people look for new jobs within ninety-one days of being hired. AI and machine learning are well suited to match people with more suitable options, whether in romantic partners (as we saw in the chapter on dating) or jobs. Similar AI tools are being used to help employees find a desirable work culture, in addition to the job itself.
- AI, algorithms, and technology platforms bring a variety of opportunities and benefits that are already shaping various aspects of work. Research suggests that an AI-based chatbot given to customer service agents increased productivity, as measured by issues resolved per hour, by 14 percent on average, with the greatest impact on novice and low-skilled workers, and minimal impact on experienced and highly skilled workers. Keep in mind that in order to explore ChatGPT's full potential, crafting good instructions that communicate our intent to an AI model will be the key to getting accurate and relevant coherent responses. We predict that this new skill known as *prompt engineering* will be an invaluable professional skill in the coming years.
- AI can mitigate and even eliminate bias in hiring. A particularly promising approach is the use of reinforcement learning to design resume-screening algorithms. These algorithms are designed to help us exploit and explore at the same time and are being shown to increase diversity in hiring while maintaining quality.

7 WELCOME TO THE AI-ENHANCED HOME

Do you use an iPhone equipped with Siri? Do you have an Amazon Echo smart speaker or other Alexa-enabled device in your home? Or maybe the little puck that responds to "Okay, Google" prompts? Several *voice-activated assistants* (also called *voice assistants*) emerged in the early 2010s and consumers snapped them up, enticed by the promise of hands-free computing and the ability to streamline daily life. Suddenly, fussing with buttons and screens felt very retro, a relic of the twentieth century—at least, that's what the ads suggested. The voice assistants had arrived, ready to lighten our loads by automating various tasks. All we had to do was ask.

Now more than a decade later, Siri, Alexa, Google Assistant, Cortana, and other voice-activated assistants have taken up residence in millions of homes. Integration with smartphones has expanded their reach even further, offering users help and information on demand. There have been some growing pains along the way, including misunderstood commands and answers that weren't exactly helpful, which comedians used as fodder for their jokes.[1] The fact that so many people can relate to these jokes is another indicator of the technology's wide acceptance and adoption. Maybe the assistants didn't deliver perfectly cooked dinners and spotless, stress-free homes where no one ever runs out of toilet paper. Nonetheless, the assistants have had quite an effect on home life, and they've shown us what's coming on the horizon.

Voice-recognition technology is quickly becoming more sophisticated. At the same time, it's being integrated with digital platforms, data, software, and an array of hardware devices. Powered by artificial intelligence and machine learning, this expanding digital ecosystem brings more convenience, comfort, and cost savings into our homes. Combined, these many innovations are bringing us closer to living the life *The Jetsons* promised.

In that classic Hanna-Barbera cartoon TV show, George Jetson and his family zip around in flying cars, have an in-home robot assistant, and use touch consoles and voice prompts to complete just about any domestic task.[2] We don't have the flying cars (yet), but today's technologies have already made many of the show's fanciful imaginings the reality. We've had robot vacuums for years, and video phone calls are now the default for many.

Thanks to an array of digitally connected smart apps and devices powered by AI, our daily homelife has had crucial upgrades improving function, comfort, and safety. But today's AI-enhanced home isn't up in the cartoon clouds like George Jetson's; it's right here on the ground and is very real.

A HOME THAT HELPS

Home is, naturally, at the center of our lives. It's where we sleep, eat, and ready ourselves before going out into the world. It's where we build lives with partners and spouses, raise kids, and entertain friends. It's where we celebrate, grieve, recuperate, and make memories.

In a more practical sense, homes require a lot of us: regular cleaning, routine maintenance, improvements as things break or wear out, the seemingly endless to-do list, not to mention paying the bills that come with all of that. Who hasn't been exhausted by the never-ending errands and chores? Who hasn't felt overwhelmed by the material stuff and the expenses and the responsibility, at least in certain moments? Maintaining a home can be stressful in the best of times.

Artificial intelligence in the *smart homes* of today and the near future centers on helping people manage stress, complete tasks, and enhance the environments in which we live. Let's take a quick tour of some AI-powered digital advancements making home life better, safer, more comfortable, and more enjoyable overall.

ENTERTAINMENT

If you're reading this book, you're likely already aware of some of the ways AI has begun to reshape our daily lives. For many people, early experiences with AI and ML came courtesy of entertainment apps such as Spotify and Netflix. Platforms like these use machine learning to analyze your behavior—in these two cases, specifically, the kinds of music you play and the shows you like to watch.

The "learning" happening on these platforms is based on data you generate while using them. What you've searched for and selected in the past, how you rate different songs or shows, how much time you spend with different genres, and other in-platform activities provide AI and ML with data these services use to build a profile of your listening and watching habits, which then opens the possibility for artificial intelligence to find other things you're likely to enjoy, whether they're new releases, older classics you may have missed, or hidden gems that have flown under the critics' radar. And because AI has access to other users on the platform, it can search for profiles with similar activity to yours and discover even more things to recommend.

Data-informed recommendations like those made by Spotify and Netflix have become common. Shopping websites, social media and social networking apps and online advertising platforms all moved in this direction years ago.

Today, as the algorithms become more advanced, they're beginning to offer even more nuanced suggestions. For example, a music platform might incorporate data about weather and time of day into its recommendations. If the algorithm learns that you tend to play high-energy songs in the morning (say, while working out) and slower, more relaxing tracks in the evening, then the platform can alter what it suggests based on this pattern. Essentially, given enough quality data, AI and ML are very good at finding the patterns in human behavior and then predicting what you're likely to want, when, and under what circumstances.

Soon your AI-enabled entertainment system will be an active, personal curator—one that finds, retrieves, and organizes content suited to your individual tastes and schedule. AI will scan television listings and note when the baseball team you watch faithfully is scheduled to play. The system will remind you of the game so you don't miss a minute of the action, or, if you prefer, will record the broadcast for you to watch later. Likewise, if a show you like releases new episodes, AI will add them to your queue automatically. When you're ready for a relaxing evening on the couch, those episodes will be waiting for you to watch.

COMFORT

Device and platform integration is driving smart home innovation even further. Imagine you've had a long day at work and a stressful evening commute. You arrive home exhausted and ready to unwind. Because your home security system is equipped with cameras and facial-recognition software, embedded artificial intelligence reads your facial expression and identifies your mood. Seeing that you're stressed, the AI-powered smart home system tells your home network to dim the lights in the living room, close the window blinds, and cues the sound system to play a low-tempo, soothing playlist from your Spotify library. You enter a calming sanctuary tailor-made just for you that allows you to de-stress and have a better evening.

If you want to experiment with this kind of home tech, there are products available on the market already. For example, AI-powered lighting systems, such as Philips Hue, use machine learning algorithms to learn user preferences and adjust lighting accordingly. These systems can also create personalized lighting scenes based on user behavior and preferences, enhancing the overall ambiance of the home.

With smart lighting, you're no longer limited to whatever bulbs you can find in the hardware store. Today's smart bulbs allow you to choose any combination of hue, temperature, and brightness that pleases you. Using the accompanying apps, you can customize the settings for different rooms or different lamps, and configure it all based on time of day, ambient lighting conditions, and so on. And once the lighting system is set up and connected to your home Internet network, you can control it all via voice command and those smart speakers.

The rapid advancements in smart home technology and device integration are transforming our living spaces into highly personalized and adaptive environments. As exemplified by AI-powered lighting systems like Philips Hue, these innovations cater to individual preferences and needs, making our homes more comfortable and responsive to our emotions and daily routines. By embracing these technologies, we not only enhance the aesthetics and ambiance of our living spaces, but also improve our overall well-being and quality of life. As these innovations continue to evolve, the smart home of the future will become an increasingly seamless and intuitive extension of our lives, making our homes true sanctuaries that adapt to and support our ever-changing needs.

SECURITY

Thanks to AI, the smart security system mentioned in the previous section does more to keep homes and people safer than do traditional security systems. Home security cameras are no longer just passive recording devices. The newest cameras, equipped with computer vision and facial-recognition software, can recognize you at your doorstep. And because your face is in the system's library of people who are approved to enter, the door unlocks automatically as you approach. You simply walk inside without having to fumble through a bag or pockets for keys. It's not quite a flying car, but it's getting close to how the Jetsons lived.

On the flip side, AI can recognize any strangers who approach the home and notify you. Using a smartphone app, you can view the camera feed from anywhere to see if the stranger is a welcome delivery driver leaving a package or someone who shouldn't be there. With a bit of input from you, AI can learn not to bother you with alerts when it's your mail carrier or the neighbor kid coming to shovel your front steps.

With AI-enabled home security you'll know when your kids arrive home from school (even when they forget to check in) and how (and what!) your pets are doing when you're away. Or you can grant a babysitter or service technician access to your home from anywhere you happen to be, all without physical keys that can be lost or stolen. And unlike touchpad locks that can be hacked, AI detects when someone tries to tamper with a lock and can contact authorities immediately.

EFFICIENCY

The AI homes of the near future will run more efficiently, conserve energy, and save you money thanks to monitoring of heating, cooling, water, and other systems. An early example of this is the Nest Thermostat. This Google product uses machine learning algorithms to learn your preferences, habits, and schedule.[3] With some initial input from you, the device optimizes heating and cooling schedules based on weather forecasts and occupancy patterns to meet your needs while minimizing waste. AI-enabled thermostats will learn our preferences and schedule, then adjust the temperature accordingly to save energy while maintaining comfort. They can also analyze your home's energy consumption patterns, identify energy "hogs" in your household, and provide recommendations to reduce usage.

The AI in such a device helps in various ways: It assumes the task of adjusting the temperature so you don't have to think about it. It goes beyond simple timers to find the settings and patterns that maximize comfort while minimizing waste. For instance, beds that allow you to set your temperature preferences and then dynamically adjust them in response to your sleep patterns are helping people with improving their sleep quality.[4] Unlike humans, AI never forgets to lower the temperature at night or when you leave the house. Soon, AI will detect changes and anomalies, like an extended lack of activity, and adjust accordingly. As a result, you'll save money while reducing your energy consumption. According to Google, "independent studies showed that [the Nest Thermostat] saved people an average of 10 percent to 12 percent on heating bills and 15 percent on cooling bills."[5]

AI is also being embedded in a new generation of home appliances, from ovens to washing machines. It may not be the sexiest of technology applications, but the benefits are significant. This kind of AI can drive down electric bills, monitor components that may be overheating, and prevent fires from endangering our families. They can also automate the process of turning off appliances and devices when not in use, further conserving energy. AI can analyze data from sensors and IoT (Internet of Things) devices to predict when home appliances may need maintenance, allowing you to address issues before they become costlier problems. Many of us know someone whose dishwasher leaked while they were on vacation, flooding their kitchen floor.

Going back to smart lighting, we will soon see AI automatically control lighting systems to optimize energy usage by adjusting brightness based on natural light levels, room occupancy, and user preferences. Some systems can learn and adapt to your habits over time. AI can even help optimize solar panel installations by predicting the best angles and positions for the panels, as well as managing energy storage and distribution based on usage patterns and weather forecasts.

CONVENIENCE

AI helps with household chores, too. Appliances such as the Roomba and other "robot" vacuum cleaners use machine learning algorithms to map out your home, learn the layout, and navigate around obstacles when it cleans. The AI in these devices also learns your preferences, such as days and times when you like it to run (or not).

AI is also becoming very good at helping people manage their digital lives and the constant influx of information coming at us through email, messaging apps, and other sources. For instance, if you're in the kitchen preparing a nice dinner, you can ask your smart speaker to check for any messages from your spouse, who has yet to arrive home. Whether the message arrives via text, email, or voicemail, your AI-powered assistant can retrieve it and read the content aloud while you continue slicing and dicing. Because your device uses natural language processing and machine learning algorithms to understand and respond to your commands, you know when your partner is running late and you can adjust the timing of the meal accordingly. You didn't even have to wash your hands to check your phone.

While you wait for your spouse and since you're in the kitchen, you notice you're running low on a few staples, including your favorite granola bars, pasta, and dish soap. All you need to do is ask your smart speaker to put these items on your shopping list. As it happens, the AI-enabled assistant has already put these items into your cart based on the pattern it learned of past order frequency. It's ready to place an order, but before it does, do you want it to add the ingredients for that recipe you downloaded to your phone earlier in the day?

Your virtual assistant app can also brief you for the next workday and set alarms for your kids to get them out of bed in the morning. Scanning your online calendar, the assistant informs you that tomorrow's off-site client meeting has been moved to 10 a.m., thirty minutes later than originally scheduled. Because of the change, the AI reports that your drive time will be faster due to lighter traffic on the route you need to take. Calculating the difference, the assistant determines you have time to take an early spin class at the gym; it suggests the idea to you and offers to reserve a spot in the class. The assistant also reminds you to bring a physical copy of an important document to the meeting. You get crucial information to help you manage your morning, all without having to log into your work account, where you're likely to get sucked into the vortex of your inbox. How convenient.

Meanwhile, companies such as Minneapolis-based Best Buy have avoided getting digitally disrupted by competitors such as Amazon by creating new lines of AI-enabled services that help elderly people-many of whom suffer from chronic disease-live longer in their own homes.[6] Best Buy's Geek Squad technicians install an array of sensors in elderly citizens' homes. Coupled with wearables and backend predictive-modeling algorithms, this AI-enabled system monitors various movement patterns and health biometrics. If any of the device's readings fall outside a defined range, both your father and his doctor are notified. The smartwatch also tracks your father's movements around his apartment. AI can recognize any unusual movements. For example, sudden movements might indicate a fall. Similarly, no movement for an extended period outside of your father's normal sleep schedule could also indicate a problem. In either case, AI can send an alert for you to check on your dad or call for emergency help if needed. All of this gives you greater peace of mind.

These examples illustrate the many ways AI-enabled technology is ready to assist you around the house. Individually, the tasks and devices described here may seem like small advantages, things people can surely live without. But added together, these smart home innovations are having a broader impact on our home lives and changing things for the better. In the AI home, mundane tasks are completed automatically, time at home is more restful, and leisure activities more enjoyable.

BUT WHAT ABOUT PRIVACY? WHAT ABOUT THE CHILDREN?

As you can tell, we're excited about what AI brings to our doorsteps, living rooms, and kitchen tables. But we also understand that the rapid pace of this technological home invasion raises questions and red flags for many. Our thoughts on some of these concerns follow.

For some people, privacy may be a concern when it comes to smart devices. They erroneously have been told that these Internet-connected appliances are listening in and collecting data, building detailed profiles of our behavior. They even worry about devices someday taking control of our lives. But are these concerns legitimate? Or do they stem more from watching futuristic movies like *The Terminator* and *Transformers*? Some AI experts say there is more hyperbole than reality in these concerns. So what do we think?

While it is essential to be vigilant, it's also important to acknowledge the extensive measures that reputable smart device manufacturers are taking to keep our information secure. Encryption, secure data storage, and regular software updates are just some of the ways these companies are working to mitigate risks. Users can further safeguard their information by using strong passwords and two-factor authentication. Taking appropriate precautions and staying informed can help ensure we enjoy the benefits of smart home technology and keeping our information secure.

We should add that what complicates the development of privacy policies and features for companies is the high degree of heterogeneity in user preferences toward data. For more than twenty years, a large body of academic work (including Anindya's papers with his coauthors) has consistently shown that there is substantial heterogeneity in a given consumer's privacy preferences and valuations across different kinds of data originating from different sources. Moreover, it is often very difficult to discern clearly what consumers' needs and wants are with respect to data privacy online. This is because what users say they want in surveys is often at odds with what their real-world behavior reveals, a phenomenon recognized in the literature as the "privacy paradox." In his first book, *TAP: Unlocking the Mobile Economy*, Anindya describes how "there is a disconnect between our understanding of what it means to be privacy-conscious in the mobile economy and the actions we are taking in the real world."[7]

Another question people often have is: "What will all of this technology do to children who grow up surrounded by it?" This is not a new concern, of course; each generation experiences (and shapes) the world differently than the one before.

In the twentieth century, people worried about the effects television, video games, headphones, and even calculators would have on kids and their development. In the 1970s, a large majority of teachers and parents worried about calculators in the classroom "for fear that their children would forget their basic math skills."[8] Here in the twenty-first century, similar concerns have been raised over kids' use of smartphones and social media, as well as personal computers and the Internet in general (not to mention ChatGPT as we discussed in a previous chapter). A quick Google search on any of these topics (including calculators) reveals that debates around the pros and cons of technological advances are ongoing and complex.

Concerns regarding the rapid spread of new technologies, while valid, often do little to halt the adoption of the technology. Even in cases where harm has been proven—such as Instagram's negative effects on the mental health of teen girls, the late-arriving proof may struggle to reverse usage trends.[9]

When it comes to AI in the home specifically, some researchers have posited that smart speakers and voice assistants could have "a long-term impact on children's social and cognitive development, specifically their empathy, compassion and critical thinking skills."[10] If kids grow up giving commands to robots, how will that affect their ability to interact with humans and process social cues?

Others see natural curiosity in kids' interest in such technology. All toddlers learn about the world by interacting with their surroundings. That includes both living things and inanimate objects. Voice-activated speakers are interesting to kids because they respond. Asking Alexa to read a story, play a favorite song, or turn on a lamp generates a response the child learns. This is not unlike manipulating a toy to learn what it does, and then repeating the action to confirm the response.[11]

There's concern, too, that chatting with Alexa and other devices could make young children bossier or promote other undesirable behaviors.[12] This is hardly a black-or-white question given the complexity of child development, family dynamics, parenting, and uses of technology in various environments. One study by Erin Beneteau and colleagues found that having a smart speaker in the home may positively affect family dynamics. In summary, they found "three forms of influence the smart speaker has on family dynamics: 1) fostering communication, 2) disrupting access, and 3) augmenting parenting."[13]

Another group of researchers found that when kids interacted with a computer "conversational agent" they learned from the experience.[14] In particular, the children who participated in the study learned an invented vocabulary word from the technology. Interestingly, the kids' subsequent interaction with parents and the researchers indicated that the children differentiated between interactions with technology and with humans.

In an article written about the study, study coauthor Alexis Hiniker stated, "I think there's a great opportunity here to develop educational experiences for conversational agents that kids can try out with their parents. There are so many conversational strategies that can help kids learn and grow and develop strong interpersonal relationships, such as labeling your feelings, using 'I' statements or standing up for others." She added, "Parents know their kid best and have a good sense of whether these sorts of things shape their own child's behavior. But I have more confidence after running this study that kids will do a good job of differentiating between devices and people."[15]

We the authors are not psychologists or pediatricians, but we are educators and parents. Our current views on AI and kids can be summarized this way: the effect of AI and smart devices on children's development is a complex and multifaceted issue that warrants careful consideration. While there are concerns about potential negative effects on social and cognitive development, research also indicates that children can differentiate between interactions with technology and humans, and that these devices may offer valuable learning experiences and even promote positive family dynamics. As educators and parents, it's essential to maintain a balanced perspective, recognizing the potential benefits of AI in the home while also staying vigilant regarding its potential drawbacks. Ultimately, the key to successfully integrating AI and smart devices into the lives of children lies in fostering open communication, using technology as a tool to supplement rather than replace human interaction, and continuously evaluating the impact of these devices on individual children and their unique developmental needs.

By fostering a healthy balance between technological interaction and traditional social experiences, we can ensure that children benefit from the educational and developmental opportunities that AI offers without compromising their emotional and social growth. Encouraging a collaborative approach that involves parents, educators, and technology developers in creating responsible and thoughtful AI experiences will play a pivotal role in shaping a future where children can thrive in a world increasingly intertwined with technology.

SMART HOMES ARE GETTING SMARTER

AI is already in our homes and has been for some time. Given that consumers have enthusiastically embraced a wide range of AI-enhanced products—from voice assistants and smart speakers to robot vacuum cleaners and automated lighting systems—we believe the trend will continue, expand, and accelerate. As AI continues to become more sophisticated, we expect to see a proliferation of innovative applications in our homes, where AI can optimize energy consumption, reduce utility costs, keep us safer, and promote eco-friendly practices. We will also continue to witness the emergence of more intelligent and personalized appliances capable of catering to individual preferences and anticipating user needs, further enhancing the comfort and convenience of our daily lives. However, with the rapid adoption of AI-enhanced products in our homes, it is crucial that we remain conscious of potential challenges and ethical considerations and the effect on human interaction. As AI becomes increasingly integrated into our living spaces, we anticipate that manufacturers and policymakers will work collaboratively to address these concerns and create a safe, responsible, and beneficial environment. As informed and AI-literate citizens, we can work with these entities to ensure a future where AI serves as a valuable tool that enhances our quality of life while preserving our fundamental values and overall well-being.

TAKEAWAYS

- The quality of daily living has had several upgrades thanks to an array of digitally connected smart apps and devices. Powered by artificial intelligence and machine learning, this expanding digital ecosystem brings more convenience, comfort, and cost savings into our homes. AI in the smart homes of today centers on helping people manage stress, complete tasks, reduce costs of living, and enhance the environments in which we live. For instance, independent studies showed that the Nest Thermostat saved people an average of 10–12 percent on heating bills and 15 percent on cooling bills.
- Virtual assistant apps can also brief us for the next workday and set alarms for our kids to get them out of bed in the morning. AI is playing a significant role in how we take care of the elderly, helping them stay in their own homes longer. Coupled with wearables and backend predictive modeling algorithms, this suite of AI-enabled tools monitors various movement patterns and health biometrics and gives alerts to caregivers.
- Consumers have enthusiastically embraced of a wide range of AI-enhanced products—from voice assistants and smart speakers to robot vacuum cleaners and automated lighting systems. We believe the trend will continue, expand, and accelerate. However, with the rapid adoption of AI-enhanced products in our homes, it is crucial that we remain conscious of potential challenges, security issues and ethical considerations, and the effect on human interaction.

8 CLIMBING THE AI SUMMIT: HOW TO BUILD AI-SAVVY ORGANIZATIONS

> The development of AI is as fundamental as the creation of the microprocessor, the personal computer, the Internet, and the mobile phone. It will change the way people work, learn, travel, get health care, and communicate with each other.
>
> —Bill Gates[1]

While the thrust of this book is to demystify the role AI for in everyday life of the average person, we would be amiss if we did not share our targeted insights on how to deploy AI successfully for managers, executives, and leaders. If this isn't relevant for you, please skip to the conclusion, which offers broader guidance for how we can go forward as a society in an AI-powered world.

Both the authors of this book have a love for the mountains going back to their adolescent years (more on this later). Hence, we will use the metaphor of climbing the AI summit to depict an organization's ability to get significant business value from deploying AI into their strategic and operational pursuits. What are the key ingredients for organizations to create business value from AI? As a teenager who went to British-style boarding school in India, Ravi was obsessed with the mighty Himalayas in general and Mt. Everest in particular. Some of his fondest childhood memories are those of summers on trekking expeditions to various base camps including Annapurna and Everest. Anindya, on the other hand, takes this to a whole new level, given his early childhood exposure to mountains like Kilimanjaro where he grew up (in Tanzania and Zambia). He is a certified high-altitude

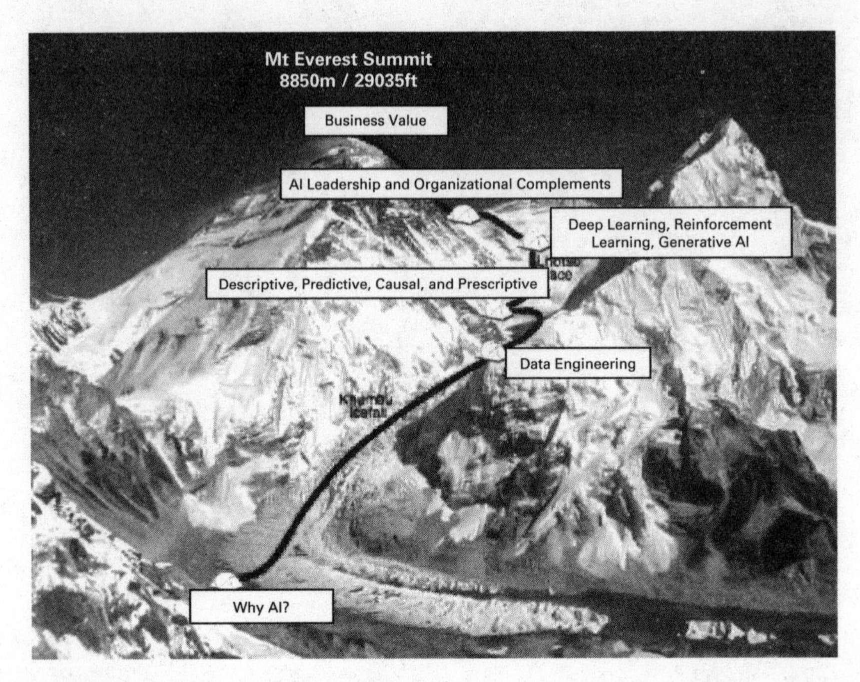

Figure 8.1
The AI summit superimposed on Mt. Everest

mountaineer with several climbs in the Andes, Cascades, Himalayas, Rockies, and Alps on his list of conquests. We'll use the classical South Face Nepalese side climb to Everest as a metaphorical learning journey for executives to create business value from AI. Please keep in mind that there are four camps in between the base and the peak of Everest. Let's begin at base camp.

BASE CAMP—SUCCESS IN AN AI-ENABLED WORLD REQUIRES LEADERS TO ADDRESS THE "WHY AI" QUESTION

The success or failure of any attempt on Everest is largely determined by the leadership displayed at base camp. Colonel John Hunt, leader of the famous 1953 expedition, did not climb Everest himself, but he led the best-run team that made the most use of favorable weather to put Tenzing Norgay and Edmund Hillary on the summit, the first people to ever stand atop it. In a similar vein, the journey from implementing AI to gaining business value does not begin if we don't recognize that leaders (national and local leaders, CEOs, and boards of institutions) face a distinct challenge in the age of AI. The leaders of tomorrow need to manage a new set of intangibles—digital technologies, digital business models, data, advanced analytics, algorithms, and now, artificial intelligence. These are very different from the classical mix of land, labor, and capital. The components of the AI economy are invisible, misunderstood, enigmatic, and therefore feared much in the same way that automobiles were greeted with hysteria when they first appeared in the early part of the last century. Leaders need to lead the charge to unbundle the key physical and digital components of their production functions and they need to find the ideal human-machine mix that utilizes the unique capabilities of humans as well as the offerings of state-of-the-art machine intelligence. In short, they need to have clarity on *why* AI needs to be deployed, what are its use cases, and how it can augment human capabilities.

How many corporate boards of non-tech companies or how many state and national legislatures are having this conversation with their executive branch? For those that are not, this is governance failure staring you in the face. Furthermore, how many board members can walk the walk and talk the talk of an AI-enabled world? How many can clearly distinguish between supervised and unsupervised machine learning or imagine how generative adversarial networks—a type of Generative AI that is useful for translating images from one domain to another (say, satellite images to maps, generating videos, etc.)—work so as to identify their use cases?

Leaders not only need to understand the art of the possible, but also to have a deeper sense of human cognition limitations that are fundamentally overcome by AI. Their main job is to constantly keep their most productive resources, namely their human capital, on the efficient frontier of what problems to solve. It's this deeper understanding that will answer the why question, which in turn will motivate investments in game-changing AI. Let's examine the first step, which takes us from the *why* to the *how*, of creating a strong foundation for turning data into a valuable business asset.

CAMP 1—IF DATA IS THE NEW OIL, DATA ENGINEERING IS THE NEW CORE COMPETENCE

Two and a half decades and trillions of dollars in IT investments, most significantly in enterprise systems such as Enterprise Resource Planning (ERP) and Customer Relationship Management (CRM), have organizations drowning in a sea of data. Time and again we hear that current corporate data yield very few actionable insights. Based on our experience we see *three patterns of confusion* about how to convert data into value.

First, each year in our data science and AI consulting engagements, both authors come across a set of corporate leaders who want to go on fishing expeditions. Undoubtedly, they have a ton of data that they are sitting on, have heard the hype around using AI to convert data into business value, and ask us to consider a project to "see if there is any value." We work with these clients to refine their thinking toward identifying a specific business problem. What keeps them up at night? What are two to three things they would like to change from the status quo? If this cannot be achieved, we defer these projects till more clarity around a situation and complication can be articulated, and a key business question identified.

Second, there is another group that underestimates the organizational politics of getting to the key features that are needed to make the analytics for AI interesting and worthwhile. These clients don't have ownership over the key data, nor do they have the power and/or influence without authority to get other stakeholders to share these data. These projects often produce underwhelming results while the most interesting problems, such as predicting whether an employee will churn, often get disproportionate benefit from machine learning's unique ability to mine the intersectionality of a multitude of factors. However, these are worthwhile projects for most organizations as they help move on from base camp of the organizational learning curve, reaching as far as camp 1 or 2, but still quite a distance from the summit.

Finally—and this get more nuanced but is all too common an occurrence—there are leaders, unlike in the first case above, who have a specific business question in mind, and who directionally have a sense of the type of AI analytics to be done. What they miss out on is the understanding of the necessary variation in the key input/predictor or output/outcome variables that make the analytics possible.

Data engineering is hard work. Having trained hundreds of analytics professionals over the last ten years at New York University and University of Minnesota Twin Cities, *data engineering* is the part of the analytics projects they dislike the most. Most professionals in the field chronically underestimate the time needed cleaning, aggregating, integrating, and organizing data to make it suitable for machine learning. Our collective analysis of more than 250 data science projects over the last ten years spanning twenty-plus industries, reveals that close to 70 percent of an analytics project's effort is consumed by data engineering. This is what happens in the base camp of a company's data science journey—all the querying, cleaning, formatting, processing of the dirty raw data to convert them into cleaned-up shiny datasets.

On Everest, the most treacherous part of the ascent is successfully navigating the Khumbu icefall. The melting ice at the relatively lower altitude and an ever-moving glacier cause large crevasses to open without warning. This is the journey from base camp to camp 1. And it is all the planning at base camp that involves setting up a foundation for a successful summit assault.

CAMP 2—LEVERAGING THE FOUR PILLARS IN THE HOUSE OF AI

If you successfully gathered the necessary equipment and made it to base camp or perhaps even advanced base camp, you are ready to make rapid progress toward the AI summit. In other words, you are now ready to go after use cases from the descriptive, predictive, causal, and prescriptive pillars. That said, more often than not, our experience working with companies and governments suggests that progress is going to be bogged down by a deadly combination of not knowing where to start and a never-ending torrent of hype headwinds. Descriptive and predictive analytics deploy unsupervised and supervised learning respectively to form the backbone of modern business applications of machine learning, or *weak AI* as we defined it. And they are classic general-purpose technologies useful across wide swaths of everyday life—not something you delegate to IT!

It's no wonder that machine learning is associated with heralding the fourth industrial revolution, occupying the same hallowed space as the steam engine, electricity, and computing associated with the three prior revolutions. The general-purpose nature of machine learning is best exemplified if we analyze where different companies in different industries started their AI and analytics journey. What were the low-hanging fruit they plucked?

The question of what to include is more art than science and requires a consultative process between the business leads and the data scientists. If you are non-tech company sitting on tons of data from decades of investments in enterprise IT systems and looking where to start, you need to use descriptive and predictive machine learning as a starting point. From working on customer segmentation in the travel industry championed by marketing to employee churn analytics in services championed by finance and HR, our personal data science stories showcase how descriptive and predictive machine learning is a powerful new general-purpose technology

Next, we address the deployment of causal analytics to deal with an increasingly complex world. It requires a test-and-learn culture that encourages experimentation using the scientific method. Its main purpose is to help leaders and managers understand the relationship between cause and effect using the long-established principles of the scientific method. While this approach is second nature in academia and has always been used by society for high-stakes decisions—can a pharmaceutical company release a new drug into the market?—it is increasingly being used for all types of decisions. It is a cornerstone of the test-and-learn culture of the "product-mindset"-based software-driven digital transformation of all companies. A variety of organizations already use some variant of the causal inference paradigm (often called "A/B testing" in the industry) to determine which ads to show, what features to deploy, or what type of incentives to provide in order to motivate users to perform an action.[2]

To be clear, as we have discussed, while tremendous business and societal value lies in using descriptive and predictive machine learning, these approaches we operate in the world of correlation. They can largely be categorized as either pattern mining of interesting associations or anomalies, or correlational mapping of inputs to outputs to predict or score data they (descriptive and predictive ML) have not encountered. They are not intended to inform us about cause and effect. Not understanding the distinction between correlation and causation is a key fault line that plagues the vast majority of the workforce, including, sadly, many workers trained in data science and analytics.

It is also imperative to note that there is nothing morally or intellectually superior about causal analytics using, say, the gold standard of double-blind randomized control trials as compared to a highly correlational predictive deep-learning model that can detect retinal disease in preemies with a degree of accuracy that is higher than a panel of the best-trained ophthalmologists.[3] These are different tools that are meant to tackle different challenges. They also spawn a vastly different type of analytics work lifecycle to execute, from data engineering to analysis to interpretation and communication of findings. Our hope is that by providing the underlying intuition of the inner workings of these approaches, leaders will become adept at picking the right approach to the problem.

As an example, one of the authors of this book was approached by a large medical device manufacturer's Business Insights team to help a fast-growing business unit understand the linkages between the time, dollars, and effort they were investing in hosting events and conferences for their physician customers and the likelihood of the physicians prescribing the focal company's device. This is a classical ROI (return on investment) marketing question that has plagued marketers for decades. In recent times, with more measurability, and seemingly more data that tracks various actions, the illusion of linking marketing actions to business outcomes has become more attractive to many. But significant pitfalls exist, and the answers lie in learning how to do causal analytics well. The president of the business unit asked a simple question, "How do we know that these prescriptions would not have been made had we not done the events? Or perhaps only a [*sic*] half of them would have been made, and could we be over-investing in events and under-investing in other marketing channels?" In the language of the data scientist, the president asked about the counterfactual.

Finally, we make the case for prescriptive analytics. Here, an organization is mixing human and machine intelligence in an optimal manner, playing to the strengths and minimizing the weaknesses of each. It uses descriptive, predictive, and casual analytics as building blocks and adds a layer of optimization over that. It's a state that's hard to achieve and rarely seen, but certainly aspirational.

CAMP 3—TACKLING MORE COMPLEX USE CASES AND DATA USING DEEP LEARNING, REINFORCEMENT LEARNING, AND GENERATIVE AI

As organizations become comfortable with developing use cases based on regular numeric/tabular data, they can expand their horizon to derive business value from richer, but more unstructured forms of data such as images, audio, video, and language. We have deliberated at length through the course of this book on the role of deep learning, reinforcement learning, and, increasingly, the use of Generative AI for such purposes.

A recent client of ours wanted help evaluating a proposal from an analytics consulting company claiming a high level of accuracy in predicting customer churn using audio data. A key challenge our client faced was determining how to rapidly evaluate the proposed model without sharing the private audio calls data that would involve getting many levels of clearances from the legal department. We guided them in the use of Generative AI techniques to *create a synthetic dataset* that resembled the statistical properties of the original audio data. This dataset was then used to evaluate the proposal of not only the vendor under consideration, but also two more of their competitors without actually sharing any private information.

In another example, a company wanted to detect a certain pattern in radiology images but had very little training data of their own. This is usually a nonstarter, because deep learning models such as convolutional neural networks require vast amounts of training examples to learn from. Our advice to this client was to use the idea of *transfer learning* to borrow, say, 95 percent of the model training done on a very large dataset by an existing open-source model, and train the last few stages of this model with their limited number of images to get a high-accuracy model despite having limited image data.

CAMP 4 TO SUMMIT—THE PATH TO BUSINESS VALUE LIES IN DEVELOPING STRONG AI LEADERSHIP AND ORGANIZATIONAL COMPLEMENTS

Camp 4 is located on the South Col pass between Mount Everest and Lhotse. The terrain here is vastly different from anything before or after it. The gusty cross winds blowing from Tibet on the North to the Khumbu region of Nepal on the South do not allow for much snow accumulation even at 26,000 feet. This is the metaphorical rough terrain that companies encounter when they try to pivot from existing judgment, intuition, and experience-based decision-making to adopting a data-driven test-and-learn culture. The journey from Camp 4 to the summit of Everest gets harder each step of the way. Austrian climber Reinhold Messner who along with Peter Habeler achieved the first ascent of Mt. Everest without oxygen in 1978 described his feeling: "In my state of spiritual abstraction, I no longer belong to myself and to my eyesight. I am nothing more than a single narrow gasping lung, floating over the mists and summits." But the achievement was epic, much in the same way British athlete Roger Bannister's cracking the four-minute mile was, because they paved the way for many others to surpass what was hitherto considered unconquerable.

The South Col equivalent on the journey to the AI summit is developing AI leadership capabilities within the organization and managing the culture change that is required to augment human decision-making with AI inputs.

For instance, the idea of running experiments to make decisions is terrifying to many organizational leaders and a growing literature on experiment-aversion is documenting this.[4] However—as Microsoft found out when they switched from the HIPPO (highest paid person's opinion) model for all product feature decisions to A/B testing—it will be worth every bit of the effort!

Like anything hard, the impetus to do this must come from leadership that is comfortable with the many components of the House of AI from chapter 1. These leaders must be familiar with the use cases of the various forms of AI and there must be a deliberate effort to educate the leadership team around the art of the possible. This will help build the organizational muscle, setting priorities that align with overall strategy, then funding projects, hiring supervisors and doers, and creating teams of business leaders, data scientists. and data engineers to get some early wins with AI. A key ingredient to get right here is your AI talent strategy.

TO SUCCEED IN AN AI-FIRST WORLD WE NEED TO CREATE AN AI-READY WORKFORCE

If the base camp is led by a visionary CEO, guided by a forward-thinking board, we need a team of climbers and Sherpas who work together to reach the summit. In our view, the Sherpas, a very proud community from the Khumbu region of Nepal, are the real heroes of most summit assaults. They scout new terrain, fix ropes so that others can benefit on their way up, and do a lot of the heavy lifting to ensure the climb's success.

Big changes are coming our way in non-tech companies for knowledge workers in the director to VP level. We refer to this as the messy middle. As AI automates more cognitive jobs—do we need that loan officer screening applications or can an algorithm trained on learning the combined expertise and wisdom of hundreds of thousands of past cases (good and bad) better serve the needs of the bank?—we'll have to recruit an army of middle managers who are problem-definers and translators. Their core competency is to identify what problems to solve and projects to fund. If a company has strong leadership that understands the *why* and a broad layer (across functions) of middle managers that is great at recognizing opportunities and defining the *what*, then we have the foundations of an AI-first corporation.

But this picture is incomplete without an army of doers, the Sherpas (the real heroes in most climbs) who are adept at *how* to apply state-of-the-art advances in AI, machine learning, and advanced analytics to bring creative solutions to problems in a wide range of industries across a wide range of functions. As former (Ravi) and current (Anindya) directors of two of the leading MS-Business Analytics programs in the world, we have first-hand experience with creating a talent pool that (1) has high business acumen—for instance, can differentiate between a balance sheet and a profit and loss statement, (2) is top-notch in data-engineering because 70 percent of time for advanced analytics projects is spent in getting the dataset cleaned, aggregated, integrated, and massaged, (3) has a deep understanding of descriptive, predictive, causal, and prescriptive analytics, and (4) undergoes extensive experiential learning to communicate the value of analytics to a broad set of stakeholders. Unlike many new graduate programs in data science, ours was not a new wrapper over a set of courses that the university already offered. Instead, they were based on our House of AI framework and had key inputs and detailed engagement from industry executives that emphasized real-world problem-solving analytics translation, storytelling and ethical and fair use of data.

Finally, in talking about creating an AI-ready workforce, we would be remiss not to tackle head-on the under-representation of women, people of color and other minorities in AI in particular and STEM in general. We will discuss positive strategies that inform us on how we can make an AI-First world align with other major forces such as diversity, equity, inclusion and purpose that are tugging at the soul of modern corporations.

AN AI-DRIVEN STRATEGY MUST PRIORITIZE FAIRNESS, ACCOUNTABILITY, AND TRANSPARENCY

Although extensive research is underway to develop fair, explainable, accountable, and transparent AI systems, the immediate threat of algorithmic bias remains. If not addressed proactively, AI can not only reproduce human biases in a mechanical manner but also potentially amplify them. It is essential to recognize the numerous sources of bias in AI-driven systems. For example, in the workforce, if AI supporters, proponents, data scientists, and researchers are predominantly male or lack adequate representation from all population segments, this inevitably leads to bias in problem-solving and solution development. Furthermore, biases can emerge from training data and algorithm construction.

To understand the issue of potential bias in AI systems, it's worth starting our journey in 2009 with the ImageNet project, co-created by Stanford University's Fei-Fei Li and Christiane Fellbaum with a mission to "map out the entire world of objects."[5] This canonical dataset was released to the

AI community as a sandbox in which to fine-tune their image classification algorithms. Most significantly, in 2012 ImageNet led to University of Toronto researchers, headed by Geoffrey Hinton, who demonstrated the potential of deep learning,[6] a technology that has since expanded to many other game-changing application areas including medicine, as discussed earlier, and the self-driving car.

Furthermore, the ImageNet project has become a fertile ground for revealing the negative consequences of AI system bias and researching ways to mitigate such issues. AI researcher Kate Crawford and artist Trevor Paglen set out to show the world the dark side of image classification using AI by running the ImageNet Roulette project.[7] People were encouraged to take a selfie and upload it to the now extinct https://imagenet-roulette.paglen.com/ site. In return the site produced labels based on a deep learning model trained on the ImageNet dataset. In a fascinating retrospective called *Excavating AI: The Politics of Images in Machine Learning Training Sets*, Crawford and Paglen discuss how as one interacts with the dataset, things get strange. A photograph of a woman smiling in a bikini is labeled a "slattern, slut, slovenly woman, trollop." A young man drinking beer is categorized as an "alcoholic, dipsomaniac, boozer, lush." And so on. Scroll through Twitter for hashtag #imagenetRoulette and you'll see people of color being labeled as "wrongdoer, offender." Imagine now if this sort of AI technology is being used to screen passengers boarding a plane or candidates for a job interview. And what if related technology, systems such as COMPAS, likewise have biases when they predict which inmate should be granted parole based on their likelihood of being re-offending?[8] ImageNet Roulette exposed the biases of not just the humans (lowly paid Amazon Turkers) who labeled the original dataset, but also the choices made by humans who set up the project that impacted how the algorithms were trained. For instance, could the choice of the hierarchical taxonomy of label categories that were selected by the ImageNet project's founders be different (it was based on WordNet, an older database of English words used for natural language processing) and if so, how would the roulette wheel spin?

Olga Russakovsky at Princeton University has been working diligently to eliminate bias from the ImageNet dataset while still keeping it useful for training object-detection AI algorithms. Her team is removing categories that may be offensive, giving power to the community to flag troublesome categories and examining how these changes impact downstream applications. She reminds us, though, of this harsh truth: "I don't think it's possible to have an unbiased human, so I don't see how we can build an unbiased AI system."[9]

TAKEAWAYS

- Deploying AI to create value is a leadership imperative for today's organizations. Executives and boards have to educate themselves on the art of the possible and deploy strategies and tactics to (1) innovate in their offerings, (2) create complementarities with existing tangible and intangible assets, and (3) keep their human capital on the efficient frontier. AI cannot be delegated to IT, or operations, or thought of as a back-office cost-cutting technology.
- Companies should not underestimate the value of data engineering in creating a strong foundation for extracting AI-enabled business value. Our experience suggests that close to 70 percent of AI project time is spent in ideating (what features to use in, say, a predictive model), cleaning, transforming, and integrating data from disparate sources within and outside the firm. Talent is scarce in both data engineering and data science, but the triathlete, a metaphor for the most valuable type of employee, who has the most value for modern firms also combines these skills with the business savvy to ask the right questions and identify the key problems to solve.
- Leaders must proactively manage the organizational and cultural change needed to go from the status quo to deploying the four pillars and three floors of the House of AI. Particular attention should be paid in developing the causal analytics pillar, as the idea of running experiments to make decisions can be worrisome for many organizational leaders. They should embrace the fact that Generative AI will play a crucial role in empowering organizations by unlocking new possibilities for creativity, innovation, and efficiency. By leveraging the power of Generative AI

models, organizations can automate complex tasks, generate realistic and personalized content, accelerate product development, improve decision-making processes, and create unique user experiences. That said, leaders have to educate themselves to evaluate machine learning models not just from the perspective of maximizing accuracy, but also their commitment to being equitable and fair in the decisions they enable.

CONCLUSION: MAKING AI WORK FOR YOU

It's the fall of 2023 in the Northern United States. For university professors, including the authors of this book, it means fresh new beginnings. As the leaves turn and a new wave of college students make their way to our respective campuses, both of us have been spending significant amount of time educating leaders around the world on making AI work for society. At a fireside chat at Mayo Clinic's RISE Conference 2023 in Minneapolis, Ravi spoke on the issue of AI and coded bias and then shared the stage with Dr. John Halamka, president of Mayo Clinic Platform (a division within Mayo Clinic tasked to create value from digital and AI tools and technologies) , who is tasked with bringing AI into his organization.[1] Halamka's key message was that for high-risk situations (e.g., a cardiologist deciding between two possible treatment pathways), current Generative AI tools, because of their proclivity to hallucinate (the Gen AI tool confidently recommended a procedure with the backing of a journal article), are a non-starter (neither the paper nor the journal exist!). In contrast, in a low-risk environment, imagine that a patient with a 5,000-page medical chart walks in and complains of leg pain, Generative AI saves that patient's doctor hours of time by summarizing the key issues, better preparing the doctor for the exam.

Soon after the RISE conference, at another event a week later, Dr. Abraham Verghese, head of internal medicine at Stanford and a renowned writer, was complaining at his book reading at the iconic Fitzgerald Theater in St. Paul about the drudgery of entering patient information into EPIC, the electronic health record system (he called it a failure of EPIC proportions). Meanwhile,

Purdue University launched a challenge for teams of analytics students to use Chat-GPT to automate that process from transcripts of medical records. Verghese is soon going to have more time to weave his sinuous stories and the world will be a better place.[2]

Anindya started that fall semester with speaking engagements in Europe and Asia at several globally renowned events. First, he delivered a keynote at the 2023 Estoril Conferences in Portugal, sharing the stage with European heads of state, Nobel laureates, entrepreneurs, Olympic athletes, explorers, journalists, adventurers, musicians, and environmentalists. The president of Portugal opened the conference and urged the audience to think deeply about the good and bad aspects of new emerging technologies. Next, Anindya spoke at the 2023 World Knowledge Forum in South Korea alongside Dr. Anthony Fauci (architect of U.S. COVID-19 policy); Sam Altman, CEO of Open AI; former Japanese prime minister Hatoyama Yukio, former U.S. Secretary of Defense James Mattis; cofounder of Apple Steve Wozniak, and Nobel laureates Paul Romer and Abhijit Banerji, among others. At both events, Anindya spoke about how AI is making the world a better place, drawing on examples from this book. He was delighted to observe tremendous enthusiasm from the audience in considering all the crucial disruptions being caused by AI. Further, he articulated what the authors have emphasized throughout this book: AI is not the problem, but rather it's the solution to a variety of societal ills that plague our times. In contrast to the conventional wisdom that is hyped by the media, AI is the solution to weeding out unacceptable and fake content from our public discourse, to maintaining civility on our digital platforms, to detecting and restricting hate speech, cyberattacks, fake accounts, and spam—and more.

Looking broadly at the market at the time of finishing this manuscript in the summer of 2023, at least by one metric, AI optimism is in the air. The S&P 500 index has risen by close to 8 percent since ChatGPT was launched and is up by close to 20 percent since the lows of October 2022. These gains are largely on the back of companies that make AI hardware (Nvidia), software (Google, Meta, and Microsoft, the latter has major investments in privately held OpenAI), infrastructure (data centers such Arista), and those that have significant AI exposure. Only time will tell whether this is irrational exuberance or a precursor to even greater ways in which AI is going to improve our lives. We wrote this book betting on the latter outcome. In contrast to big media's incessant emphasis on a fear-based narrative, we believe that AI in many ways is propelled by the fuel of at least four major technological revolutions that have preceded it—the Internet, mobile, cloud, and data. We are not alone in lamenting the panic and hysteria over the potential bad outcomes of AI. Marc Andreessen, the man who gave the world the web browser, calls this fear and paranoia an irrational moral panic, one that is clouding our ability to confront actual concerns.[3]

We believe that modern AI is going to power breakthrough technologies that will solve the grand challenges of today concerning climate change, disease, wealth and income inequality, and our polarized society. In the preceding chapters we showcased examples of how artificial intelligence is radically changing the way we go about detecting disease, inventing cures, and equalizing the opportunities for education (we showcased the possibility of a personalized tutor for every child on the planet), and for finding meaningful work and relationships. The record-breaking adoption of ChatGPT—we regularly use it as a co-pilot today to augment our executive teaching—is indicative of our hope-based narrative.

We took you on a journey that showcases how modern AI enhances virtually all the essential aspects of your everyday life—who you connect with, who you date, your home, your health and wellness, your education, and your work life. AI does so by making predictions (e.g., is a student likely to graduate in four years), recommendations (e.g., here is a potential date you are probably going to be attracted to, or a book or a song or a movie you are going to like), generating new data (e.g., creating textual content or an image or a short video for your Instagram Reels post), or by learning a series of actions to perform a task that would otherwise take you time and energy (e.g., a robot optimally cleaning your house's floor while you are away). In the process of showcasing how AI is embedded in your day-to-day life, we have, hopefully, also given you reasonable agency into the inner workings of AI. You now know the importance of data engineering (the ability to integrate, cleanse, and curate a variety of structured and unstructured data), and how it feeds into the descriptive, predictive, prescriptive, and causal pillars of the House of AI. You should be able to relate to the importance of deep learning in taking us beyond the realm of tabular data to the world of text, images, audio, and video; as well as to the importance of algorithmic advances such as "attention mechanisms" in representing contextual information in a way that we can model the entire span of human language (e.g., GPT-4). This is the AI-enriched world we are living in, and we hope that it is more approachable to you than when you began to read the book.

This leaves us with the all-important challenge of how to *thrive* in the age of AI: how to best negate the likelihood of a dystopian future where AI overtakes the human capacity to do good, increase innovation and productivity, and ultimately the quality of life of all citizens on this planet. There is a lot of talk about how best to regulate AI given all the power it potentially has to cause harm to society. There is little doubt in our mind that any AI regulation enacted must be multifaceted in nature, protecting citizens and nation-states from harm while simultaneously encouraging innovation, entrepreneurship, and productivity growth. AI regulation has to balance the interests of the state, firms, and individuals with the aim of improving the human condition and reducing existing disparities. Our hope in writing this book is that by developing agency and understanding in the conversation around AI, citizens will become more informed and actively participate in the political process of shaping regulation. Bereft of such participation, we might end up with regulations crafted by politicians influenced by lobbyists that sound well intended but are not fully thought out and are loaded with unintended consequences. Europe's General Data Protection Regulation (GDPR) comes to mind where the compliance costs are such that they favor large businesses over smaller entrepreneurial ventures, thereby limiting innovation and future benefits to society)[4] Plenty of academic research has documented the negative consequences of GDPR.[5]

If we understand the power of modern AI as coming from a highly thorough (and yes, mathematical) representation of expressed, and therefore extant, language, art, code, decisions, and more broadly, human communication, we can formulate how to go forward and innovate on top of it. As the headline for an opinion piece by *New York Times* columnist David Brooks put it, "In the Age of A.I., Major in Being Human."[6] The skills that are going to be at a premium in the age of AI are those that Brooks mentions in his article—personal voice, childlike creativity, situational awareness, and the ability to connect the dots with an unusual worldview.

When we think about what makes up the constitution of people who are universally admired at work or in life, we get a blueprint of how we can double down on "being human" and can *thrive in the age of AI.* Chances are such individuals are incredibly generous with their time and resources. They are creators of stronger bonds between neighbors, teams, and families. They may go out of their way to nurture relationships, and yet they are ethical, not afraid to not afraid to tell it like it is, and not afraid to call out a friend or family member if they are going astray. These are folks who have seen and traveled the world. Collectively, their lived experiences span cultures through which they understand the unequal distribution of wealth, income, resources, and opportunities that make up our planet today. They are constantly striving to reduce these disparities, understanding that there are multiple paths to success. They have high cultural savvy, can read the tea leaves at a cocktail party or a roundtable negotiation, and exhibit a high emotional quotient along with IQ. They are honest and reliable in their actions, always seeking solutions to problems, and even better, they excel in finding the right problems to solve. They are excellent storytellers who work the arc of the story to imprint their ideas that are backed up by evidence. At the same time, they are evidence-seeking when it comes to consuming news and information. They are lifelong learners who are curious and flexible enough to be open to new perspectives. And yet, they are brave enough to take a stand, voice uncomfortable opinions, and create safe-enough spaces such that diverse viewpoints can be heard without the need for consensus.

We think of AI as a strong complement to human intelligence. The infinitely wise Buddha once said, "Thousands of candles can be lit from a single candle, and the life of the candle will not be shortened." Thriving in a world increasingly influenced by artificial intelligence involves understanding its potential, fostering the necessary skills to engage with it, and utilizing its capabilities to enhance personal and professional life. These include the following:

Educate yourself: Start by understanding what AI is, how it works, and its potential impacts. There's a wealth of resources available, from books such as this to online courses to podcasts and articles. Becoming knowledgeable about AI can help dispel fears that arise from misunderstandings and equip you to make informed decisions about AI usage in your life.

Embrace lifelong learning: As AI evolves, so too will the skills needed to thrive in an AI-driven world. Embrace the concept of lifelong learning, continually updating your knowledge and skills. AI is good at automating repetitive tasks, so it is important to develop skills that are not easily automated and are complementary to AI. Focus on skills such as critical thinking, creativity, and emotional intelligence. These *soft skills* are difficult for AI to replicate and are highly valued in many professions.

Engage in ethical discussions: Join and encourage discussions about the ethical implications of AI. Issues such as data integrity, job displacement, and the potential for AI bias are important to consider. By engaging in these discussions at a public level, you can contribute to the development of regulations and societal norms that guide AI usage. It is important to become a responsible user of artificial intelligence. AI can be used for good or not-so-good purposes, so it is important to use it responsibly. This means being aware of the potential risks of AI and using it in ways that benefit society.

Prepare for change: While embracing AI, maintain a healthy level of skepticism of the information you consume. Not every AI application will be beneficial or without risk. It's important to consider the credibility of the source and the purpose of the AI tool. AI is still in its nascent stage, and there is a lot of misinformation out there about it. Be sure to fact-check the information you read and watch, and don't be afraid to ask questions. AI will continue to evolve and change many aspects of daily life. Being open and adaptable to these changes will be key to thriving in an AI-driven world.

Acknowledgments

This book was inspired by the dichotomy we saw between the narrative in the media about AI taking our jobs and leading us to the end of human civilization versus our lived experiences of the last twenty years creating and witnessing the positive impact of artificial intelligence, machine learning, and advanced analytics. We are indebted to our family, friends, colleagues, coauthors, students, and industry partners who encouraged and motivated us to broadly share our perspectives on this topic.

First and foremost, we want to acknowledge the pivotal role of our dear families and friends.

Ravi: I would like to dedicate this book to my late father, Dr. Jawahar Singh Bapna, a pharmacologist who wrote *The Complete Family Medicine Book*, which demystifies modern medicine for laypeople. The book is in its twelfth edition and has sold more than one million copies.[1] My father inspired me to pursue a life of research, teaching, and scholarship, and to do so in a positive way that brings out the best in everybody. Nothing in my career would be possible without the love, encouragement, and support of my wife, Sofia, and our daughter, Mehek. Among other things, they are my in-house Reviewer 1 and Reviewer 2 and inspire me every day to do my best. So do family who are friends (Mummy, Nana and Nani, Mama and Mami, Masi and Masosa, Swinky, Ajit Chacha, Sunita Ma, Shakuntala Mamisa), and friends who are family: the Mayo 86–87 batch "Dosts" (Durgesh and Digsa, Moti, and Shelja; Nandi and Chinky; Manish and Kavita; Sharad and Malika; Manji Kanota and Sandhya Bhabisa; Pashu and

Daffy; and Munna and Jayshree), Deepti, Shomik, and Renu; Sunil Dutta and Sheryl Hoye; Amit and Madhu; and the Manipal 83 friends (Karna, Shishir, Sethia, Fandu, Bhatia, Arjun, Premal, Saira, Bob and Shweta, and Anu Nadella). Thanks for always having my back.

Anindya: I give a special shoutout to my wife, Deepti, for her love and invaluable support in helping me chase my professional and personal dreams, and to our daughter, Ananya, for enriching my life in so many ways, including by becoming my mountaineering partner. I thank my parents, Aloke and Anindita, and father-in-law, Satish, for their love and unwavering support. I thank other members of my family including Alaka, Amitabh, Rahul, Priya, Arshia, and Anoushka for their support. Thank you, Alex and Shahana, Sofia, Ashish and Lavangana, Sunil and Atrayee, Sandeep and Namrata and Uday, for generously sharing your time, company, and wisdom. From the IIMC Class of 1998, I thank Shinjit, Nilesh, Prashant, Ritesh, Pankaj, and several wingmates from H2 (Old Hostel) for generously celebrating my wins with me over the years.

Many of the ideas in the book come from our collaboration with a wonderful set of colleagues, coauthors, and legions of graduate students and executives who we have coached. We would like to thank dozens of folks in these respective communities. This book would not have been possible without our shared experiences in harnessing the power of AI, ML, and advanced analytics in so many contexts over the last two decades.

Anindya: I am grateful to my recent Ph.D. students who have essentially become part of my extended family and have enriched my life in many ways. They include Panos Adamopolous, Gordon Burtch, Jason Chan, Hongxian Huang, Eric Kwon, Heesung (Andrew) Lee, Beibei Li, Prasanna Parasurama, Chenshuo Sun, Wally Wang, Vilma Todri, and Yuqian Xu. I thank all my coauthors and academic collaborators from the past twenty years and would particularly like to give a shoutout to recent collaborators including Maxime Cohen, Runshan Fu, Xitong Guo, Hanna Halaburda, Raghu Iyengar, Dongwon Lee, Xiao Liu, Xueming Luo, Meghanath Macha, Raveesh Mayya, Dominik Molitor, Wonseok Oh, Michael Pinedo, Martin Spann, Daniel Sokol, Alex Tuzhilin, Sriram Venkataraman, and Pearl Yu for their friendship and insights. I also thank the wonderful NYU

Stern community, especially those starting from the MSBA Class of 2018 and the EMBA and TRIUM communities, for creating many evenings filled with wonderful memories in the United State and various other parts of the world.

Ravi: I benefited tremendously in all aspects of my academic and life journey from my advisors, Professors Alok Gupta and Paulo Goes, and from early mentors, Professors Jim Marsden and Robert Garfinkel at the University of Connecticut, and Vallabh "Samba" Sambamurthy, who is now the dean of the Wisconsin School of Business. Ideas in the book have been chiseled over long evenings laced with fine wine and conversation in snowy Minnesota with Sri and Aks Zaheer, and with Svjetlana Madzar and Miki Hondzo. Much gratitude to them. In the same vein, I would like to give a shoutout to my coauthors, Gedas Adomavicious, Gord Burtch, Jason Chan, Sandeep Gangarapu, Ram Gopal, Jae Jung, Nishtha Langer, Ed McFowland, Amit Mehra, Jui Ramaprasad, Tianshu Sun, Gautam Ray, Sarah Rice, Galit Shmueli, Akhmed Umyarov, and Meizi Zhou for their insights and brilliance.

We are both grateful to several industry colleagues and friends for their support and collaboration.

Anindya: I am especially indebted to Daniel Fischel for welcoming me at Compass Lexecon, which changed the trajectory of my career. Working as a testifying expert on the highest-profile and globally watched antitrust and privacy litigation matters has massively enhanced the breadth and depth of my knowledge and has been a truly life-changing experience for me and my family. I benefited a great deal from working with Todd Kendall, Niall MacMenamin, and Nancy Wang, as well as many other colleagues at Compass Lexecon. I thank my past colleagues at Cornerstone Research and Analysis Group including Rahul Guha, Avigail Kifer, Shankar (Sean) Iyer, and Rebecca Kirk Fair. From other spheres of the corporate world, I deeply thank Akshay Chaturvedi, Julia Cho, Saugata Gupta, Rajesh Jain, Jonathan Kopnick, Ram Sellaratnam, and Wally Wang for all the exciting opportunities they have enabled for me.

Ravi: From the industry, I would like to thank Nandan Nilekani for always being generous with his time and of course for showing the world the value of population scale digital public infrastructures such as Aadhar and UPI. I am also very thankful to Manik Gupta, Joe Golden, Jonathan Hersh, Mike Martiny, Tan Moorthy, Ashok Reddy, Christian Rudder, Ellen Trader, and Pramod Varma for their invaluable industry perspective and collaboration.

We benefited a great deal from helpful conversations about life and topics related to this book with Indranil Bardhan, Ramnath Chellappa, Brett Danaher, Carlos Fernandez, Pedro Ferreira, Bin Gu, Alok Gupta, Khim Yong Goh, Avi Goldfarb, Byung Cho Kim, Ramayya Krishnan, Xitong Li, Amit Mehra , Balaji Padmanabhan, Gal Oestreicher-Singer, V. Sambamurthy, Paramvir Singh, Prasanna Tambe, Catherine Tucker, Sunil Wattal, and Rong Zheng. During the time he spent at New York University, Carnegie Mellon, and Wharton, Anindya has special gratitude in this regard for Mor Armony, David Bell, Eric Bradlow, Gerard Cachon, Tulin Erdem, Pete Fader, Elizabeth Morrison, Tridas Mukhopadhyay, Uday Rajan, Rob Seamans, Raghu Sundararam, Batia Wiesenfeld, Russ Winer, and Eitan Zemel for their advice and support at critical times in his career.

Many people gave us great feedback on our manuscript. Panos Adamopoulos, Sheryl Hoye, Narayan Ramachandran, Mindy Tsai, and Dr. Anjali Bhagra read early drafts and gave critical feedback that helped sharpen our message. We are indebted to them for their time and generosity. Kathleen Caruso and Julia Collins from the MIT Press as well as Lee Thomas and Mindee Forman provided excellent editorial assistance. The book would not have taken the shape it did without their help. Last but certainly not least, we are grateful to Catherine Woods at MIT Press for believing in the thesis of this book.

Notes

CHAPTER 1

1. Matthew Rosenberg, Nicholas Confessore, and Carole Cadwalladr, "How Trump Consultants Exploited the Facebook Data of Millions," *New York Times*, March 17, 2018, https://www.nytimes.com/2018/03/17/us/politics/cambridge-analytica-trump-campaign.html.

2. Stuart A. Thompson and Charlie Warzel, "Opinion | Twelve Million Phones, One Dataset, Zero Privacy," *New York Times*, December 19, 2019, https://www.nytimes.com/interactive/2019/12/19/opinion/location-tracking-cell-phone.html.

3. Frank Pasquale and Gianclaudio Malgieri, "Opinion | If You Don't Trust A.I. Yet, You're Not Wrong," *New York Times*, July 30, 2021, https://www.nytimes.com/2021/07/30/opinion/artificial-intelligence-european-union.html.

4. Fabrizio Dell'Acqua, Edward McFowland, Ethan R. Mollick, Hila Lifshitz-Assaf, Katherine Kellogg, Saran Rajendran, Lisa Krayer, François Candelon, and Karim R. Lakhani, "Navigating the Jagged Technological Frontier: Field Experimental Evidence of the Effects of AI on Knowledge Worker Productivity and Quality," Harvard Business School Technology & Operations Mgt. Unit Working Paper No. 24-013, September 15, 2023, https://ssrn.com/abstract=4573321.

5. Julia Dressel and Hany Farid, "The Accuracy, Fairness, and Limits of Predicting Recidivism," *Science Advances* 4, no. 1 (January 17, 2018), https://doi.org/10.1126/sciadv.aao5580. This article documents, for instance, that the COMPAS algorithm used to predict recidivism for the purposes of criminal sentencing makes systematically different mistakes for Black and white defendants: Black defendants who did not recidivate were incorrectly predicted to reoffend at a rate of 44.9 percent, nearly twice as high as their white counterparts at 23.5 percent, and white defendants who did recidivate were incorrectly predicted to not reoffend at a rate of 47.7 percent, nearly twice as high as their Black counterparts at 28.0 percent.

6. Helen Johnson, "The (Im)Proper Meshing of the Corporate Media and the Military-Industrial Complex," *Miscellany News*, May 13, 2021, https://miscellanynews.org/2021/05/13/opinions/the-improper-meshing-of-the-corporate-media-and-the-military-industrial-complex/.

7. Cathy O'Neil, *Weapons of Math Destruction: How Big Data Increases Inequality and Threatens Democracy* (New York: Crown, 2017); Sufiya Umoja Noble, *Algorithms of Oppression: How Search Engines Reinforce Racism* (New York: NYU Press, 2018); Matthew Gault, "A Dystopia Where AI Runs U.S. Healthcare and Asks Patients to Die," *CYBER*, August 18, 2022, https://shows.acast.com/cyber/episodes/a-dystopia-where-ai-runs-us-healthcare-and-asks-patients-to-.
8. James Vincent, "'Godfathers of AI' honored with Turing Award, the Nobel Prize of Computing," *The Verge*, March 27, 2019, https://www.theverge.com/2019/3/27/18280665/ai-godfathers-turing-award-2018-yoshua-bengio-geoffrey-hinton-yann-lecun.
9. Zoe Kleinman and Chris Vallance, "AI 'Godfather' Geoffrey Hinton Warns of Dangers as He Quits Google," *BBC News*, May 2, 2023, https://www.bbc.com/news/world-us-canada-65452940.
10. Yann LeCun [@ylecun]: "This is absolutely correct. The most common reaction by AI researchers to these prophecies of doom is face palming," Twitter, May 4, 2023, 8:05 a.m., https://twitter.com/ylecun/status/1654125161300520967.
11. Elizabeth Gibney, "The Scant Science behind Cambridge Analytica's Controversial Marketing Techniques," *Nature*, March 29, 2018, https://doi.org/10.1038/d41586-018-03880-4.
12. Dean Eckles, Brett A. Gordon, and Garrett M. Johnson, "Field Studies of Psychologically Targeted Ads Face Threats to Internal Validity," *Proceedings of the National Academy of Sciences of the United States of America* 115, no. 23 (May 18, 2018), https://doi.org/10.1073/pnas.1805363115.
13. Taylor McNeil, "Did Cambridge Analytica Sway the Election?" *Tufts Now*, May 17, 2018, https://now.tufts.edu/2018/05/17/did-cambridge-analytica-sway-election.
14. "Compass Lexecon Client Meta (Formerly Facebook) Prevails in Consumer Privacy Suit Related to Cambridge Analytica," June 7, 2023, https://www.compasslexecon.com/cases/compass-lexecon-client-meta-formerly-facebook-prevails-in-consumer-privacy-suit-related-to-cambridge-analytica/.
15. Android, "Emergency Location Service," n.d., https://www.android.com/safety/emergency-help/emergency-location-service/.
16. Anindya Ghose, Beibei Li, Meghanath Macha, Chenshuo Sun, and Natasha Zhang Foutz, "Trading Privacy for the Greater Social Good: How Did America React During COVID-19?," Social Science Research Network working paper, January 1, 2022, https://doi.org/10.2139/ssrn.3624069.
17. Statements issued by FCC Chairman Wheeler and Commissioners Clyburn and Rosenworcel on *Third Further Notice of Proposed Rulemaking* (NPRM) in the Matter of Wireless E911 Location Accuracy Requirements (FCC 14–13), Washington, DC: Federal Communications Commission, February 21, 2014, 15, https://www.documentcloud.org/documents/2195636-fcc-third-nprm-february-2014.html#document/p15.
18. Garth H. Rauscher, Jenna Khan, Michael L. Berbaum, and Emily F. Conant, "Potentially Missed Detection with Screening Mammography: Does the Quality of Radiologist's

Interpretation Vary by Patient Socioeconomic Advantage/Disadvantage?," *Annals of Epidemiology* 23, no. 4 (April 2013): 210–214, https://doi.org/10.1016/j.annepidem.2013.01.006.

19. Brad N. Greenwood, Rachel R. Hardeman, Laura Huang, and Aaron Sojourner, "Physician–Patient Racial Concordance and Disparities in Birthing Mortality for Newborns," *PNAS (Proceedings of the National Academy of Sciences of the United States of America)* 117, no. 35 (August 17, 2020): 21194–21200, https://doi.org/10.1073/pnas.1913405117.

20. A common approach computes Euclidean distance based on the Pythagorean theorem—$a^2 + b^2 = c^2$—that comes to us from Pythagoras, a Greek philosopher born around 570 BC. See "Pythagorean Theorem | Definition & History," *Encyclopedia Britannica*, March 10, 2005, https://www.britannica.com/science/Pythagorean-theorem.

21. More sophisticated anomaly-detection algorithms exist. We teach these to our analytics graduate students, but they are not the subject of this book. Here, we take the shortest path to the mechanics behind these algorithms, acknowledging that more sophisticated approaches exist.

22. At the risk of repetition, more sophisticated predictive classification algorithms exist. We teach these to our analytics graduate students, but they are not the subject of this book. Here, we take the shortest path to the intuition behind these algorithms, acknowledging that more sophisticated approaches exist.

23. Nils Strodthoff and Claas Strodthoff, "Detecting and Interpreting Myocardial Infarction Using Fully Convolutional Neural Networks," *Physiological Measurement* 40, no. 1 (January 15, 2019): 015001, https://doi.org/10.1088/1361-6579/aaf34d.

24. Zhaoqi Cheng, Dokyun Lee, and Prasanna Tambe, "InnoVAE: Generative AI for Understanding Patents and Innovation," Social Science Research Network working paper, March 1, 2022, https://doi.org/10.2139/ssrn.3868599.

25. OpenAI, "GPT-4 Technical Report," *ArXiv*, (2023), accessed June 27, 2023, https://arxiv.org/abs/2303.08774.

26. Shunyuan Zhang, Dokyun Lee, Param Vir Singh, and Kannan Srinivasan, "What Makes a Good Image? Airbnb Demand Analytics Leveraging Interpretable Image Features," *Management Science* 68, no. 8 (August 1, 2022): 5644–5666, https://doi.org/10.1287/mnsc.2021.4175.

27. Jeffrey Dastin, "Amazon Scraps Secret AI Recruiting Tool That Showed Bias against Women," *Reuters*, October 10, 2018, https://www.reuters.com/article/us-amazon-com-jobs-automation-insight/amazon-scraps-secret-ai-recruiting-tool-that-showed-bias-against-women-idUSKCN1MK08G.

28. Danielle Li, Lindsey R. Raymond, and Peter Bergman, "Hiring as Exploration," *National Bureau of Economic Research* Working Paper No. w27736, August 1, 2020, https://doi.org/10.3386/w27736.

29. Jae U. Jung, Ravi Bapna, Jui Ramaprasad, and Akhmed Umyarov, "Love Unshackled: Identifying the Effect of Mobile App Adoption in Online Dating," *MIS Quarterly* 43, no. 1 (January 1, 2019): 47–72, https://doi.org/10.25300/misq/2019/14289.

30. Edward McFowland, Sandeep Gangarapu, Ravi Bapna, and Tianshu Sun, "A Prescriptive Analytics Framework for Optimal Policy Deployment Using Heterogeneous Treatment Effects," *MIS Quarterly* 45, no. 4 (October 14, 2021): 1807–1832, https://doi.org/10.25300/misq/2021/15684.

CHAPTER 2

1. David Kushner, "Recruiting Women to Online Dating Was a Challenge," *The Atlantic*, April 10, 2019, https://www.theatlantic.com/technology/archive/2019/04/how-matchcom-digitized-dating/586603/.
2. Sara Murphy, "88% of You Will Swipe Right for This," *Refinery29*, October 3, 2015, https://www.refinery29.com/en-us/2015/10/95118/grammar-online-dating.
3. Michael Mager, "Could Bad Grammar Mean a Lonely Valentine's Day for Dating Hopefuls?," *Grammarly* (blog), February 1, 2019, https://www.grammarly.com/blog/could-bad-grammar-mean-a-lonely-valentines-day-for-dating-hopefuls/.
4. Tim Harford, "Online Dating? Swipe Left," *Financial Times*, February 12, 2016, https://www.ft.com/content/b1a82ed2-8e34-11e5-8be4-3506bf20cc2b. Quoting Michael Norton, psychologist at Harvard Business School.
5. Nancy Jo Sales, "Tinder Is the Night," *Vanity Fair | the Complete Archive*, September 1, 2015, https://archive.vanityfair.com/article/2015/9/tinder-is-the-night. Quoting Justin Garcia of Indiana University's Kinsey Institute for Research in Sex, Gender, and Reproduction.
6. Michael J. Rosenfeld and Reuben J. Thomas, "Searching for a Mate: The Rise of the Internet as a Social Intermediary," *American Sociological Review* 77, no. 4 (June 13, 2012): 523–547 at 526, https://doi.org/10.1177/0003122412448050.
7. Michael J. Rosenfeld, Reuben J. Thomas, and Sonia Hausen, "Disintermediating Your Friends: How Online Dating in the United States Displaces Other Ways of Meeting," *Proceedings of the National Academy of Sciences* 116, no. 36 (August 20, 2019): 17753–17758 at 17754, https://doi.org/10.1073/pnas.1908630116.
8. Rosenfeld, Thomas, and Hausen, "Disintermediating Your Friends": "Individuals might not want to share their dating preferences and activities with their mother or with their friends. Active brokerage of romantic partnerships by a family member or friend would depend on the broker knowing what both individuals desire in a partner. Taking advantage of Facebook to find friends of friends for romantic matches (i.e., passive brokerage by friends) might expose dating habits and choices to too broad an audience. Dating perfect strangers encountered online is potentially more discreet than dating a friend's friend. A corollary to the discretion inherent in online dating is that the online precursor to face-to-face meeting inserts a layer of physical distance that can have benefits for safety." See also Brett P. Kennedy, "A History of the Digital Self: The Evolution of Online Dating," *Psychology Today*, September 22, 2010, https://www.psychologytoday.com/us/blog/the-digital-self/201009/history-the-digital-self-the-evolution-online-dating: "Anonymity allowed people to be themselves or a creative

version of thereof. Chat rooms allowed people to take risks and be bold in the expression of their intimate selves."

9. Jennifer L. Gibbs, Nicole B. Ellison, and Rebecca D. Heino, "Self-Presentation in Online Personals," *Communication Research* 33, no. 2 (April 1, 2006): 152–177 at 152, https://doi.org/10.1177/0093650205285368.

10. "The online dating sites have the potential to improve their matching algorithms through data analysis, experiments, and machine learning over time." In Michael J. Rosenfeld, Reuben Thomas, and Sonia Hausen, "Disintermediating Your Friends: How Online Dating in the United States Displaces Other Ways of Meeting," *Proceedings of the National Academy of Sciences* 116, no. 36 (August 20, 2019): 17753–17758 at 17754, https://doi.org/10.1073/pnas.1908630116.

11. eharmony Editorial Team, "eharmony's 32 Dimensions of Compatibility Explained," *eharmony*, November 4, 2021, https://www.eharmony.co.uk/dating-advice/using-eharmony/32-dimensions-compatibility-explained.

12. David Gelles, "Inside Match.com: It's All about the Algorithm," *Slate Magazine*, July 30, 2011, https://slate.com/human-interest/2011/07/inside-match-com-it-s-all-about-the-algorithm.html.

13. For example, Adam Joinson, "Causes and Implications of Disinhibited Behavior on the Internet," in *Psychology and the Internet: Intrapersonal, Interpersonal, and Transpersonal Implications*, ed. Jayne Gackenbach (San Diego: Academic Press, 1998), 43–60; John R. Suler, "The Online Disinhibition Effect," *Cyberpsychology & Behavior* 7, no. 3 (June 1, 2004): 321–326, https://doi.org/10.1089/1094931041291295.

14. Tina M. Harris and Pamela J. Kalbfleisch, "Interracial Dating: The Implications of Race for Initiating a Romantic Relationship," *Howard Journal of Communications* 11, no. 1 (January 1, 2000): 49–64, https://doi.org/10.1080/106461700246715; John E. Pachankis and Marvin R. Goldfried, "Social Anxiety in Young Gay Men," *Journal of Anxiety Disorders* 20, no. 8 (January 1, 2006): 996–1015, https://doi.org/10.1016/j.janxdis.2006.01.001.

15. Ravi Bapna, Jui Ramaprasad, Galit Shmueli, and Akhmed Umyarov, "One-Way Mirrors in Online Dating: A Randomized Field Experiment," *Management Science* 62, no. 11 (November 1, 2016): 3100–3122, https://doi.org/10.1287/mnsc.2015.2301.

16. Bapna et al., "One-Way Mirrors in Online Dating," 3101; emphasis in original.

17. "In a traditional setting, people log onto dating websites from their computers. Because dating apps run on smartphones, users can use dating apps anywhere at any time—similar to making a phone call." In Lik Sam Chan, "Who Uses Dating Apps? Exploring the Relationships among Trust, Sensation-Seeking, Smartphone Use, and the Intent to Use Dating Apps Based on the Integrative Model," *Computers in Human Behavior* 72 (July 1, 2017): 246–258 at 247, https://doi.org/10.1016/j.chb.2017.02.053.

18. "The unique features of dating apps set mobile dating apart from online dating in general. More precisely, dating apps are likely to increase the salience of dating among users as users

can receive 'push notifications' informing them about new matches and/or conversations throughout the day. The geolocation functionality of dating apps also allows users to search for someone in close proximity, which may facilitate actual offline meetings with matches." In Sindy R. Sumter and Laura Vandenbosch, "Dating Gone Mobile: Demographic and Personality-Based Correlates of Using Smartphone-Based Dating Applications among Emerging Adults," *New Media & Society* 21, no. 3 (October 20, 2018): 655–673 at 656, https://doi.org/10.1177/1461444818804773.

See also Ranzini, Giulia, and Christoph Lutz, "Love at First Swipe? Explaining Tinder Self-Presentation and Motives," *Mobile Media and Communication* 5, no. 1 (September 16, 2016): 80–101 at 82, https://doi.org/10.1177/2050157916664559: "The portability of smartphones and tablets permits the use of Tinder in a variety of locations, from private to semipublic and public spaces. By contrast, the use of traditional desktop-based dating sites is mostly restricted to private spaces. Moreover, the availability affordance of mobile media increases the spontaneity and use frequency of the app. The locatability affordance enables matching, texting, and meeting with users in close proximity—one of the key aspects of Tinder . . . this affordance seems to invite more social uses than traditional dating, for example by making swiping and gossiping about profiles a fun activity among friends."

19. Chris Fox, "10 Years of Grindr: A Rocky Relationship," *BBC News*, March 25, 2019, https://www.bbc.com/news/technology-47668951.

20. Kara Carlson, "What Is Bumble and How It Grew into an Industry Power—and How It Expects to Keep Growing," *Austin American-Statesman*, February 12, 2021, https://www.statesman.com/story/business/2021/02/11/how-bumble-became-dating-app-powerhouse/6724959002/.

21. "Following the success of the highly popular dating apps Tinder and Grindr, various new dating apps, such as Happn and Bumble, emerged. In addition, several traditional dating websites also developed their own apps (e.g., OkCupid)." In Sindy R. Sumter and Laura Vandenbosch, "Dating Gone Mobile: Demographic and Personality-Based Correlates of Using Smartphone-Based Dating Applications among Emerging Adults," *New Media & Society* 21, no. 3 (October 20, 2018): 655–673 at 665, https://doi.org/10.1177/1461444818804773.

22. Niloofar Abolfathi, "Dating Disruption—How Tinder Gamified an Industry," *MIT Sloan Management Review* Reprint #61325, February 13, 2020, at 2, https://sloanreview.mit.edu/article/dating-disruption-how-tinder-gamified-an-industry/; emphasis in original.

23. Jeana H. Frost, Zoe Chance, Michael I. Norton, and Dan Ariely, "People Are Experience Goods: Improving Online Dating with Virtual Dates," *Journal of Interactive Marketing* 22, no. 1 (February 1, 2008): 51–61, https://doi.org/10.1002/dir.20107; emphasis in original.

24. Niloofar Abolfathi, "Dating Disruption—How Tinder Gamified an Industry," *MIT Sloan Management Review* Reprint #61325, February 13, 2020, https://sloanreview.mit.edu/article/dating-disruption-how-tinder-gamified-an-industry/.

25. Aaron Smith, "15% of American Adults Use Online Dating Sites or Mobile Apps," Pew Research Center: Internet, Science & Tech, February 11, 2016, https://www.pewresearch

.org/internet/2016/02/11/15-percent-of-american-adults-have-used-online-dating-sites-or-mobile-dating-apps/.

26. "Tinder grew so fast by seeding its app with college students in the US, assuming they'd be an influential group because, as Mateen puts it: 'as someone who's younger in high school, you want to be a college kid. And a lot of adults are envious of college kids too.'" In Stuart Dredge, "Tinder: The 'Painfully Honest' Dating App with Wider Social Ambitions," *Guardian*, February 24, 2014, https://www.theguardian.com/technology/2014/feb/24/tinder-dating-app-social-networks.

27. "Beginning in 1997, online dating starts to gain prominence. This coincides with the rise of Web 2.0 technologies, such as dynamic webpages based on databases rather than static html pages." In Bernie Hogan, nai li, and William H. Dutton, "A Global Shift in the Social Relationships of Networked Individuals: Meeting and Dating Online Comes of Age," Social Science Research Network working paper, February 14, 2011, at 10, https://papers.ssrn.com/sol3/papers.cfm?abstract_id=1763884.

28. Michael J. Rosenfeld and Reuben Thomas, "Searching for a Mate," *American Sociological Review* 77, no. 4 (June 13, 2012): 523–547 at 544, https://doi.org/10.1177/0003122412448050. The authors theorized that young people had more alternative means to find partners and thus had less need for Internet dating.

29. "Perhaps the most powerful protection that small entrant firms enjoy as they build the emerging markets for disruptive technologies is that they are doing something that it simply does not make sense for the established leaders to do. Despite their endowments in technology, brand names, manufacturing prowess, management experience, distribution muscle, and just plain cash, successful companies populated by good managers have a genuinely hard time doing what does not fit their model for how to make money. Because disruptive technologies rarely make sense during the years when investing in them is most important, conventional managerial wisdom at established firms constitutes an entry and mobility barrier that entrepreneurs and investors can bank on." In Clayton M. Christiansen, *The Innovator's Dilemma* (Boston: Harvard Business School Press, 1997), 228, https://www.hbs.edu/faculty/Pages/item.aspx?num=46.

30. "Two key factors underpinned Tinder's sudden success: focusing on young adults, an overlooked market segment; and introducing new gamelike features, such as swiping and variable rewards, which altered the user experience and reduced consumption barriers in that specific segment." In Niloofar Abolfathi, "Dating Disruption—How Tinder Gamified an Industry," *MIT Sloan Management Review* Reprint #61325, February 13, 2020, at 1–2, https://sloanreview.mit.edu/article/dating-disruption-how-tinder-gamified-an-industry/.

31. "Tinder's new algorithm ranks users according to how successfully they match with others. . . . This provides incentive to strategically play the game, generating reputation-based authenticity, as defined by the algorithm." In Stefanie Duguay, "Dressing up Tinderella: Interrogating Authenticity Claims on the Mobile Dating App Tinder," *Information, Communication & Society* 20, no. 3 (March 30, 2016): 351–367, https://doi.org/10.1080/1369118x.2016.1168471.

32. Takuma Kakehi, "Extra-Gamified: Why Are Some Apps So Satisfying?," *Medium*, March 21, 2019, https://uxdesign.cc/extra-gamified-why-are-some-apps-so-satisfying-7ae8df998394.

33. Anil Isisag, "Mobile Dating Apps and the Intensive Marketization of Dating: Gamification as a Marketizing Apparatus," in *NA—Advances in Consumer Research*, vol. 47, ed. Rajesh Bagchi, Lauren Block, and Leonard Lee (Duluth, MN: Association for Consumer Research, 2019), 135–141 at 136.

34. Niloofar Abolfathi, "Dating Disruption—How Tinder Gamified an Industry," *MIT Sloan Management Review* Reprint #61325, February 13, 2020, at 4, https://sloanreview.mit.edu/article/dating-disruption-how-tinder-gamified-an-industry/.

35. Janelle Ward, "What Are You Doing on Tinder? Impression Management on a Matchmaking Mobile App," *Information, Communication & Society* 20, no. 11 (November 6, 2016): 1644–1659 at 1649–1650, https://doi.org/10.1080/1369118X.2016.1252412.

36. Laura Stampler, "Inside Tinder: Meet the Guys Who Turned Dating into an Addiction," *Time*, February 6, 2014, https://time.com/4837/tinder-meet-the-guys-who-turned-dating-into-an-addiction/.

37. Scott Hurff, *Designing Products People Love: How Great Designers Create Successful Products*, O'Reilly Media, Inc., 2015, https://www.oreilly.com/library/view/designing-products-people/9781491923696/.

38. Eric Johnson, "Swiping on Tinder Is Addictive. That's Partly Because It Was Inspired by an Experiment That 'Turned Pigeons into Gamblers.'" *Vox*, September 19, 2018, https://www.vox.com/2018/9/19/17877004/nancy-jo-sales-swiped-hbo-documentary-tinder-dating-app-addictive-pigeon-kara-swisher-decode-podcast. Quoting journalist Nancy Jo Sales on her HBO documentary *Swiped*:

    It's all related to an experiment done by B. F. Skinner, the very controversial—and some say very insidious—sociologist whose whole work was a lot about controlling behavior. "Oh look. Look what we can do. Look what we can make people do," or make pigeons do. He did this experiment—and then I went and had my archival researcher go and find the footage of the experiment, it's in the film—where Skinner, this kind of, you know, some people would say evil genius, is surrounded by cages with pigeons. And this is what Tinder is. It's like the pigeon becomes a gambler, because when he pecks and gets food, he gets bored, so he peck-peck-pecks, he doesn't know when he's gonna get the food. He might get it, he might not. That's the whole swiping mechanism. You swipe, you might get a match, you might not. And then you're just like excited to play the game. . . . You just keep playing the game. Because it's like, "Am I gonna match?" or "I might not match." And people will match like . . . we talked to people who would just match, match, match, match, match, and it was really just about the swiping and the matching more . . . it became more about that.

39. Scott Hurff, *Designing Products People Love: How Great Designers Create Successful Products*, O'Reilly Media, Inc., 2015, https://www.oreilly.com/library/view/designing-products-people/9781491923696/.

40. Monica Anderson and Emily A. Vogels, "Young Women Often Face Sexual Harassment Online—Including on Dating Sites and Apps," *Pew Research Center*, March 6, 2020,

https://www.pewresearch.org/fact-tank/2020/03/06/young-women-often-face-sexual-harassment-online-including-on-dating-sites-and-apps/.

41. Stefanie Duguay, "Dressing Up Tinderella: Interrogating Authenticity Claims on the Mobile Dating App Tinder," *Information, Communication & Society* 20, no. 3 (March 30, 2016): 351–367 at 356, https://doi.org/10.1080/1369118X.2016.1168471.

42. Duguay, "Dressing Up Tinderella," at 359: "The inability to post photos from a phone's camera shapes authenticity claims by constraining visual biographical references to those preestablished with acquaintances on Facebook and Instagram."

43. Thomas Barrie, "How Whitney Wolfe Herd Created Bumble, the $13 Billion Dating App That Will Save the Internet," *British GQ*, May 17, 2021, https://www.gq-magazine.co.uk/lifestyle/article/whitney-wolfe-herd-interview-2021.

44. Barrie, "How Whitney Wolfe Herd Created Bumble."

45. Charlotte Alter/Austin, "How Whitney Wolfe Herd Turned a Vision of a Better Internet into a Billion-Dollar Brand," *Time*, March 19, 2021, https://time.com/5947727/whitney-wolfe-herd-bumble/.

46. Thedatingverse, "Virtual Reality Date Coaching," accessed September 22, 2022, https://www.thedatingverse.com/.

47. HBO, "*We Met in Virtual Reality*, official website for the HBO series, n.d., https://www.hbo.com/movies/we-met-in-virtual-reality.

CHAPTER 3

1. Makena Kelly, "Inside Nextdoor's 'Karen Problem,'" *The Verge*, June 8, 2020, https://www.theverge.com/21283993/nextdoor-app-racism-community-moderation-guidance-protests.

2. Bobby Allyn, "It's 'Our Fault': Nextdoor CEO Takes Blame for Deleting of Black Lives Matter Posts," *NPR*, July 1, 2020, https://www.npr.org/2020/07/01/886147665/it-s-our-fault-nextdoor-ceo-takes-blame-for-censorship-of-black-lives-matter-pos.

3. Alison Van Houten, "Time100 Most Influential Companies of 2022; Nextdoor: Prompting Kindness," *Time*, March 30, 2022, https://time.com/collection/time100-companies-2022/6159411/nextdoor-leaders/.

4. Nextdoor, "Nextdoor Launches New Neighbor Features to Increase Transparency and Encourage Constructive Conversations," press release, May 3, 2022, https://about.nextdoor.com/press-releases/nextdoor-launches-new-neighbor-features-to-increase-transparency-and-encourage-constructive-conversations/.

5. Mike Schuster and Kuldip K. Paliwal, "Bidirectional Recurrent Neural Networks," *IEEE Transactions on Signal Processing* 45, no. 11 (January 1, 1997): 2673–2681, https://doi.org/10.1109/78.650093.

6. Sepp Hochreiter and Jürgen Schmidhuber, "Long Short-Term Memory," *Neural Computation* 9, no. 8 (November 1, 1997): 1735–1780, https://doi.org/10.1162/neco.1997.9.8.1735.

7. Jaipur Literature Festival, "Home," September 16, 2013, accessed November 15, 2022, https://jaipurliteraturefestival.org/.

8. Hector Yee and Bar Ifrach, "Aerosolve: Machine Learning for Humans," *The Airbnb Tech Blog*, in *Medium*, June 4, 2015, https://medium.com/airbnb-engineering/aerosolve-machine-learning-for-humans-55efcf602665.

9. Shunyuan Zhang, Nitin Mehta, Param Vir Singh, and Kannan Srinivasan, "Frontiers: Can an Artificial Intelligence Algorithm Mitigate Racial Economic Inequality? An Analysis in the Context of Airbnb," *Marketing Science* 40, no. 5 (September 1, 2021): 813–820, https://doi.org/10.1287/mksc.2021.1295.

10. Jennifer Skeem and Christopher Lowenkamp. "Using Algorithms to Address Trade-Offs Inherent in Predicting Recidivism," *Behavioral Sciences & the Law* 38, 3 (2020), 259–278, https://doi.org/10.1002/bsl.2465.

11. Julia Angwin, Jeff Larson, Surya Mattu, and Lauren Kirchner, "Machine Bias," *ProPublica*, May 23, 2016, https://www.propublica.org/article/machine-bias-risk-assessments-in-criminal-sentencing.

12. David M. Blei, Andrew Y. Ng, and Michael I. Jordan, "Latent Dirichlet Allocation," *Journal of Machine Learning Research* 3 (January 2003): 993–1022, https://jmlr.csail.mit.edu/papers/v3/blei03a.html.

13. Michelle Du, "Discovering and Classifying In-App Message Intent at Airbnb," *Medium*, January 22, 2019, https://medium.com/airbnb-engineering/discovering-and-classifying-in-app-message-intent-at-airbnb-6a55f5400a0c.

14. Airbnb, "A Letter from Co-Founder Joe Gebbia," *Airbnb Newsroom*, July 21, 2022, https://news.airbnb.com/a-letter-from-co-founder-joe-gebbia/.

15. Brian Chesky [@bchesky], "My letter to the Airbnb team about @jgebbia. [picture 4]," Twitter, July 21, 2021, 11:39 p.m., https://twitter.com/bchesky/status/1550354758266863616/photo/4.

16. "A Letter from Co-Founder Joe Gebbia," *Airbnb Newsroom*, July 21, 2022, https://news.airbnb.com/a-letter-from-co-founder-joe-gebbia/.

17. Rob Walker, "Airbnb Pits Neighbor Against Neighbor in Tourist-Friendly New Orleans," *New York Times*, March 13, 2016, https://www.nytimes.com/2016/03/06/business/airbnb-pits-neighbor-against-neighbor-in-tourist-friendly-new-orleans.html.

18. Makarand Mody, Courtney Suess, and Tarik Dogru, "Does Airbnb Impact Non-Hosting Residents' Quality of Life? Comparing Media Discourse with Empirical Evidence," *Tourism Management Perspectives* 39 (July 1, 2021): 100853, https://doi.org/10.1016/j.tmp.2021.100853.

19. Louis Bouchard, "How Uber Uses AI to Serve You Better," *Louis Bouchard* (blog), May 21, 2022, https://www.louisbouchard.ai/uber-deepeta/.

20. John Koetsier, "Uber Might Be the First AI-First Company, Which Is Why They 'Don't Even Think About It Anymore,'" *Forbes*, August 22, 2018, https://www.forbes.com/sites

/johnkoetsier/2018/08/22/uber-might-be-the-first-ai-first-company-which-is-why-they-dont-even-think-about-it-anymore/.

21. "How Lyft Is Using AI to Keep Customers Happy (VB Live)," *VentureBeat*, May 24, 2018, https://venturebeat.com/ai/how-lyfts-using-ai-to-keep-customers-happy-vb-live/.

22. Jay Caspian Kang, "The Boy King of YouTube," *New York Times Magazine*, January 5, 2022, https://www.nytimes.com/2022/01/05/magazine/ryan-kaji-youtube.html.

23. Arun Sundararajan, *The Sharing Economy: The End of Employment and the Rise of Crowd-Based Capitalism* (Cambridge, MA: MIT Press, 2017).

24. Paul Resnick and Richard J. Zeckhauser, "Trust among Strangers in Internet Transactions: Empirical Analysis of eBay's Reputation System," in *The Economics of the Internet and E-Commerce (Advances in Applied Microeconomics, Vol. 11)*, ed. Michael R. Baye (Bingley, UK: Emerald Publishing Limited, 2002), 127–157.

25. Jorge Mejia and Chris Parker, "When Transparency Fails: Bias and Financial Incentives in Ridesharing Platforms," *Management Science* 67, no. 1 (January 1, 2021): 166–184, https://doi.org/10.1287/mnsc.2019.3525.

26. Bobby Allyn, Bobby. "Uber Fires Drivers Based On 'Racially Biased' Star Rating System, Lawsuit Claims," *NPR*, October 26, 2020, https://www.npr.org/2020/10/26/927851281/uber-fires-drivers-based-on-racially-biased-star-rating-system-lawsuit-claims.

27. Megan Rose Dickey, "Study Says Uber and Lyft Have Racial Discrimination Problems," *TechCrunch*, October 31, 2016, https://techcrunch.com/2016/10/31/study-uber-and-lyft-racial-discrimination/.

28. Joshua Brustein, "Discrimination Runs Rampant throughout the Gig Economy, Study Finds," *Mashable*, November 22, 2016, https://mashable.com/article/gig-economy-discrimination-bloomberg.

29. Benjamin Edelman, Michael Luca, and Dan Svirsky, "Racial Discrimination in the Sharing Economy: Evidence from a Field Experiment," *American Economic Journal: Applied Economics* 9, no. 2 (April 1, 2017): 1–22, https://doi.org/10.1257/app.20160213.

30. Bastian Jaeger and Willem W. A. Sleegers, "Racial Disparities in the Sharing Economy: Evidence from More than 100,000 Airbnb Hosts across 14 Countries," *Journal of the Association for Consumer Research* 8, no. 1 (January 1, 2023): 33–46, https://doi.org/10.1086/722700.

31. Anagha Srikanth, "Despite Changes, LGBTQ+ and Racial Discrimination Persists in Uber, Lyft," *The Hill*, July 30, 2020, https://thehill.com/changing-america/respect/equality/509817-despite-changes-lgbtq-and-racial-discrimination-persists-in/.

32. Brian Chesky, "Fighting Discrimination and Creating a World Where Anyone Can Belong Anywhere," *Airbnb* (blog), September 8, 2016, https://news.airbnb.com/fighting-discrimination-and-making-airbnb-more-diverse/.

33. Juliet Bennett Rylah, "Airbnb's New Experiment to Reduce Racism," *The Hustle*, January 7, 2022, https://thehustle.co/01072022-airbnb-racial-bias/.

34. Jan Overgoor, "Experiments at Airbnb," *The Airbnb Tech Blog*, in *Medium*, May 27, 2014, https://medium.com/airbnb-engineering/experiments-at-airbnb-e2db3abf39e7.

35. "Airbnb Report on Travel and Living," *Airbnb*, May 2021, https://news.airbnb.com/wp-content/uploads/sites/4/2021/05/Airbnb-Report-on-Travel-Living.pdf.

CHAPTER 4

1. March of Dimes, "Nowhere to Go: Maternity Care Deserts across the U.S. (2022 Report)," *March of Dimes*, October 2022, https://www.marchofdimes.org/maternity-care-deserts-report.

2. John Patrick Pullen, "Why Professional Athletes Love This Fitness Band," *Time*, April 18, 2017, https://time.com/4744459/whoop-strap-fitness-tracker-band/.

3. Wei Wang, Gordon Blackburn, Milind Y. Desai, Dermot Phelan, Lauren Gillinov, Penny L. Houghtaling, and Marc Gillinov, "Accuracy of Wrist-Worn Heart Rate Monitors," *JAMA Cardiology* 2, no. 1 (January 1, 2017): 104, https://doi.org/10.1001/jamacardio.2016.3340.

4. Summer R. Jasinski, Shon Rowan, David M Presby, Elizabeth Claydon, and Emily R Capodilupo, "Wearable-Derived Maternal Heart Rate Variability as a Novel Digital Biomarker of Preterm Birth," *MedRxiv (Cold Spring Harbor Laboratory)*, November 5, 2022, https://doi.org/10.1101/2022.11.04.22281959.

5. WHOOP, "Understanding Pregnancy with Groundbreaking New Research & Pregnancy Coaching," November 2, 2022, https://www.whoop.com/thelocker/understanding-every-stage-of-pregnancy-with-new-pregnancy-coaching/.

6. WHOOP, "Improving Heart Rate Accuracy: Your WHOOP is Getting Smarter!," October 10, 2017, https://www.whoop.com/the-locker/improving-heart-rate-accuracy-whoop-getting-smarter/.

7. Michael B. Phillips, Jason Beach, R. Michael Cathey, Jake Lockert, and William Satterfield. "Reliability and Validity of the Hexoskin Telemetry Shirt," *Journal of Sport and Human Performance* 5, no. 2 (September 2017), https://doi.org/10.12922/jshp.v5i2.8. See also Alyssa Nolte, "Adventure Technology Could Make Extreme Climbing (a Little) Easier," *Now*, November 12, 2019, https://now.northropgrumman.com/adventure-technology-could-make-extreme-climbing-a-little-easier/.

8. impacX, "Smart Water Packaging Solutions by Water.io," Water.io, July 15, 2021, https://impacx.io/water-io/.

9. Eric Wicklund, "mHealth in Space: Not Just Science Fiction Any More," *mHealthIntelligence*, December 23, 2015, https://mhealthintelligence.com/news/mhealth-in-space-not-just-science-fiction-any-more.

10. Grand View Research, *mHealth Market Size, Share & Trends Analysis Report by Component (Wearables, mHealth Apps), by Services (Monitoring Services, Diagnosis Services), by*

Participants, by Region, and Segment Forecasts, 2024–2030, February 17, 2022, https://www.grandviewresearch.com/industry-analysis/mhealth-market.

11. Siddhartha Mukherjee, "A.I. versus M.D.: What Happens When Diagnosis Is Automated?," *New Yorker*, April 3, 2017, https://www.newyorker.com/magazine/2017/04/03/ai-versus-md.
12. Adam Satariano and Cade Metz,"How A.I. Is Being Used to Detect Cancer That Doctors Miss," *New York Times*, March 5, 2023, https://www.nytimes.com/2023/03/05/technology/artificial-intelligence-breast-cancer-detection.html.
13. US Census Bureau, "Computer and Internet Use in the United States: 2018," Census.gov, April 21, 2021, https://www.census.gov/newsroom/press-releases/2021/computer-internet-use.html.
14. Pew Research Center, "Mobile Fact Sheet," April 7, 2023, https://www.pewresearch.org/internet/fact-sheet/mobile/#who-owns-cellphones-and-smartphone.
15. Laura Silver, "Smartphone Ownership Is Growing Rapidly Around the World, but Not Always Equally," Pew Research Center's Global Attitudes Project, February 5, 2019, https://www.pewresearch.org/global/2019/02/05/smartphone-ownership-is-growing-rapidly-around-the-world-but-not-always-equally/.
16. Brian Heater, "Apple Offers a Deeper Dive into Crash Detection," *TechCrunch*, October 10, 2022, https://techcrunch.com/2022/10/10/apple-offers-a-deeper-dive-into-crash-detection/.
17. Apple Support, "Take an ECG with the ECG App on Apple Watch," September 12, 2022, https://support.apple.com/en-us/HT208955.
18. EMR is an acronym for Electronic Medical Record and EHR is an acronym for Electronic Health Record.
19. Anindya Ghose, Xitong Guo, Beibei Li, and Yuanyuan Dang, "Empowering Patients Using Smart Mobile Health Platforms: Evidence of a Randomized Field Experiment," *MIS Quarterly* 46, no. 1 (February 15, 2022): 151–192, https://doi.org/10.25300/misq/2022/16201.
20. Ghose et al., "Empowering Patients."
21. Stanley P. Rowland, J. Mark FitzGerald, Tord Holme, Jade Powell, and Alison H. McGregor, "What Is the Clinical Value of MHealth for Patients?," *npj Digital Medicine* 3, no. 4 (January 13, 2020), https://doi.org/10.1038/s41746-019-0206-x.
22. For example, Cristian Pop-Eleches, Harsha Thirumurthy, James Habyarimana, Joshua Graff Zivin, Markus Goldstein, Damien De Walque, Leslie D. MacKeen, et al., "Mobile Phone Technologies Improve Adherence to Antiretroviral Treatment in a Resource-Limited Setting: A Randomized Controlled Trial of Text Message Reminders," *AIDS* 25, no. 6 (March 27, 2011): 825–834, https://doi.org/10.1097/qad.0b013e32834380c1.
23. Hang Yin, Daniel R. Neuspiel, Ian M. Paul, Wayne J. Franklin, Joel S. Tieder, Terry A. Adirim, Francisco J. Alvarez, et al., "Preventing Home Medication Administration Errors," *Pediatrics* 148, no. 6 (December 1, 2021), https://doi.org/10.1542/peds.2021-054666.

24. Yin et al., "Preventing Home Medication Administration Errors."

25. Jenn, "ER 8.5, Start All Over Again: Everyone's Having a Horrible Day (Well, Maybe Not Rachel)," *'90s Flashback*, March 9, 2021, https://90sflashback.wordpress.com/2021/03/09/er-8-5-start-all-over-again-everyones-having-a-horrible-day-well-maybe-not-rachel/.

26. La Princess C.Brewer, Sarah M. Jenkins, Sharonne N. Hayes, Ashok Kumbamu, Clarence F. Jones, Lora E. Burke, Lisa A. Cooper, and Christi A. Patten, "Community-Based, Cluster-Randomized Pilot Trial of a Cardiovascular Mobile Health Intervention: Preliminary Findings of the FAITH! Trial," *Circulation* 146, no. 3 (July 19, 2022): 175–190, https://doi.org/10.1161/circulationaha.122.059046.

27. Stanley P. Rowland, J. Mark FitzGerald, Tord Holme, Jade Powell, and Alison H. McGregor, "What Is the Clinical Value of mHealth for Patients?," *npj Digital Medicine* 3, no. 4 (January 13, 2020), https://doi.org/10.1038/s41746-019-0206-x.

28. World Health Organization (WHO), "COVID-19 Pandemic Triggers 25% Increase in Prevalence of Anxiety and Depression Worldwide," press release, March 2, 2022, https://www.who.int/news/item/02-03-2022-covid-19-pandemic-triggers-25-increase-in-prevalence-of-anxiety-and-depression-worldwide.

29. "Access to Care Data 2022: Access to Care Ranking 2022," *Mental Health America*, 2022, https://mhanational.org/issues/2022/mental-health-america-access-care-data.

30. Kashyap Kompella, "The Pros and Cons of Using AI-Based Mental Health Tools," *Information Today, Inc.*, September 27, 2022, https://newsbreaks.infotoday.com/NewsBreaks/The-Pros-and-Cons-of-Using-AIBased-Mental-Health-Tools-155090.asp.

31. Anindya Ghose, Xitong Guo, Beibei Li, and Yuanyuan Dang, "Empowering Patients Using Smart Mobile Health Platforms: Evidence of a Randomized Field Experiment," *MIS Quarterly* 46, no. 1 (February 15, 2022): 151–192, https://doi.org/10.25300/misq/2022/16201.

32. Thomas Kramer, Suri Spolter-Weisfeld, and Maneesh Thakkar, "The Effect of Cultural Orientation on Consumer Responses to Personalization," *Marketing Science* 26, no. 2 (March 1, 2007): 246–258, https://doi.org/10.1287/mksc.1060.0223.

33. Rhonda Hadi and Ana Valenzuela, "Good Vibrations: Consumer Responses to Technology-Mediated Haptic Feedback," *Journal of Consumer Research* 47, no. 2 (August 2020): 256–271, https://academic.oup.com/jcr/article-abstract/47/2/256/5559276.

34. Che-Wei Liu, Guodong Gao, and Ritu Agarwal, "Reciprocity or Self-Interest? Leveraging Digital Social Connections for Healthy Behavior," *Management Information Systems Quarterly* 46, no. 1 (January 20, 2022): 261–298, https://doi.org/10.25300/misq/2022/16177.

35. Alex Perry, "Google Finally Ends Support for the Original Google Glass," *Mashable*, December 7, 2019, https://mashable.com/article/google-glass-explorer-edition-final-update.

36. Saturday Night Live, "Weekend Update: Randall Meeks," September 25, 2013, https://www.youtube.com/watch?v=5Uz3cwHT0S0.

37. Comedy Central, “Glass Half Empty,” *The Daily Show*, June 16, 2014, https://www.youtube.com/watch?v=ClvI9fZaz6M.

38. Google Glass Help, “Final Software Update for Glass Explorer Edition,” n.d., https://support.google.com/glass/answer/9649198?hl=en&ref_topic=3063354.

39. Glass Early Access Program, “Discover Glass Enterprise Edition,” n.d., https://www.google.com/glass/start/.

40. Glass Early Access Program, “Case Studies,” n.d., https://www.google.com/glass/case-studies/.

41. Erin Digitale, “Google Glass Helps Kids with Autism Read Facial Expressions,” *Stanford Medicine News Center*, August 2, 2018, https://med.stanford.edu/news/all-news/2018/08/google-glass-helps-kids-with-autism-read-facial-expressions.html.

42. Stanford Medicine, “Behavioral Therapy Sessions for Your Home,” The Autism Glass Project at Stanford Medicine!, n.d., https://autismglass.stanford.edu/.

43. Center for Devices and Radiological Health, “Artificial Intelligence and Machine Learning (AI/ML)-Enabled Medical Devices,” *U.S. Food & Drug Administration*, October 5, 2022, https://www.fda.gov/medical-devices/software-medical-device-samd/artificial-intelligence-and-machine-learning-aiml-enabled-medical-devices.

44. Elise Reuter, “5 Takeaways from the FDA's List of AI-Enabled Medical Devices,” *MedTech Dive*, November 7, 2022, https://www.medtechdive.com/news/FDA-AI-ML-medical-devices-5-takeaways/635908/.

45. Varun Gulshan, Lily Peng, Marc Coram, Martin C. Stumpe, Derek Wu, Arunachalam Narayanaswamy, Subhashini Venugopalan, et al., “Development and Validation of a Deep Learning Algorithm for Detection of Diabetic Retinopathy in Retinal Fundus Photographs,” *JAMA* 316, no. 22 (December 13, 2016): 2402, https://doi.org/10.1001/jama.2016.17216.

46. University of Minnesota Carlson School of Management, “Optum: Speaking the Language of Data Analytics,” Executive Education, n.d., https://carlsonschool.umn.edu/executive-education/custom-programs/success-stories/optum.

47. National Health Care Anti-Fraud Association (NHCAA), “The Challenge of Health Care Fraud,” n.d., https://www.nhcaa.org/tools-insights/about-health-care-fraud/the-challenge-of-health-care-fraud/.

48. Eri Sugiura and Leo Lewis, “AI Is Giving Insurers Godlike Powers, Says Sompo Chief,” *Financial Times*, November 13, 2022, https://www.ft.com/content/a3372e1a-d43c-403e-97e5-449b50d51b87.

49. Ewen Callaway, “‘It Will Change Everything’: DeepMind's AI Makes Gigantic Leap in Solving Protein Structures,” *Nature* 588, no. 7837 (November 30, 2020): 203–204, https://doi.org/10.1038/d41586-020-03348-4.

50. Ewen Callaway, “‘The Entire Protein Universe’: AI Predicts Shape of Nearly Every Known Protein,” *Nature* 608, no. 7921 (July 28, 2022): 15–16, https://doi.org/10.1038/d41586-022-02083-2.

51. Jamie Smyth, "Biotech Begins Human Trials with Drug Discovered Using AI," *Financial Times*, October 31, 2022, https://www.ft.com/content/0006ae3f-7064-4aa6-98cd-8912f544acc5.

52. Merck, "Merck and Moderna Announce Exercise of Option by Merck for Joint Development and Commercialization of Investigational Personalized Cancer Vaccine—Merck.Com," press release, October 12, 2022, https://www.merck.com/news/merck-and-moderna-announce-exercise-of-option-by-merck-for-joint-development-and-commercialization-of-investigational-personalized-cancer-vaccine/.

CHAPTER 5

1. OpenAI, "About," September 2, 2020, https://openai.com/about/.

2. Mark Lieberman, "What Is ChatGPT and How Is It Used in Education?," *Education Week*, January 27, 2023, https://www.edweek.org/technology/what-is-chatgpt-and-how-is-it-used-in-education/2023/01.

3. Greg Brockman [@gdb], "ChatGPT just crossed 1 million users; it's been 5 days since launch," Twitter, December 5, 2022, 1:32 a.m., https://twitter.com/gdb/status/1599683104142430208

4. Jacob Stern, "Five Chats to Help You Understand ChatGPT," *The Atlantic*, December 16, 2022, https://www.theatlantic.com/technology/archive/2022/12/openai-chatgpt-chatbot-messages/672411/; Kevin Roose, "The Brilliance and Weirdness of ChatGPT," *New York Times*, December 6, 2022, https://www.nytimes.com/2022/12/05/technology/chatgpt-ai-twitter.html.

5. Kathy Hirsh-Pasek and Elias Blinkoff, "ChatGPT: Educational Friend or Foe?," *Brookings*, June 9, 2023, https://www.brookings.edu/blog/education-plus-development/2023/01/09/chatgpt-educational-friend-or-foe/.

6. Kevin Roose, "The Brilliance and Weirdness of ChatGPT," *New York Times*, December 6, 2022, https://www.nytimes.com/2022/12/05/technology/chatgpt-ai-twitter.html.

7. Gary Marcus, "AI Platforms Like ChatGPT Are Easy to Use but Also Potentially Dangerous," *Scientific American*, December 19, 2022, https://www.scientificamerican.com/article/ai-platforms-like-chatgpt-are-easy-to-use-but-also-potentially-dangerous/.

8. Marcus, "AI Platforms Like ChatGPT Are Easy to Use."

9. Jacob Stern, "Five Chats to Help You Understand ChatGPT," *The Atlantic*, December 16, 2022, https://www.theatlantic.com/technology/archive/2022/12/openai-chatgpt-chatbot-messages/672411/.

10. Ian Bogost, "ChatGPT Is Dumber Than You Think," *The Atlantic*, December 16, 2022, https://www.theatlantic.com/technology/archive/2022/12/chatgpt-openai-artificial-intelligence-writing-ethics/672386/.

11. Ethan Mollick, "ChatGPT Is a Tipping Point for AI," *Harvard Business Review*, December 14, 2022, https://hbr.org/2022/12/chatgpt-is-a-tipping-point-for-ai.

12. Kevin Roose, "Don't Ban ChatGPT in Schools. Teach with It," *New York Times*, January 13, 2023, https://www.nytimes.com/2023/01/12/technology/chatgpt-schools-teachers.html.

13. Kathy Hirsh-Pasek and Elias Blinkoff, "ChatGPT: Educational Friend or Foe?," *Brookings*, June 9, 2023, https://www.brookings.edu/blog/education-plus-development/2023/01/09/chatgpt-educational-friend-or-foe/.

14. Ashish Vaswani, Noam Shazeer, Jakob Uszkoreit, Llion Jones, Aidan N. Gomez, Łukasz Kaiser, and Illia Polosukhin, "Attention Is All You Need," in *Advances in Neural Information Processing Systems 30 (NIPS 2017)*, ed. I. Guyon, U. Von Luxburg, S. Bengio, H. Wallach, R. Fergus, S. Vishwanathan, and R. Garnett, https://papers.nips.cc/paper_files/paper/2017/hash/3f5ee243547dee91fbd053c1c4a845aa-Abstract.html.

15. OpenAI, "GPT-4," March 14, 2023, https://openai.com/research/gpt-4.

16. OpenAI, "GPT-4."

17. OpenAI, "GPT-4."

18. Jesus Rodriguez and K. Se, "Edge 256: The Architecture and Methods Powering ChatGPT," *TheSequence*, December 29, 2022, https://thesequence.substack.com/p/edge-256-the-architecture-and-methods.

19. Rodriguez and Se, "Edge 256."

20. Ian Bogost, "ChatGPT Is Dumber Than You Think," *The Atlantic*, December 16, 2022, https://www.theatlantic.com/technology/archive/2022/12/chatgpt-openai-artificial-intelligence-writing-ethics/672386/.

21. Peter Coy, "Opinion | ChatGPT Can't Do My Job Quite Yet," *New York Times*, December 16, 2022, https://www.nytimes.com/2022/12/16/opinion/chatgpt-artificial-intelligence-skill-job.html.

22. Ethan Mollick, "ChatGPT Is a Tipping Point for AI," *Harvard Business Review*, December 14, 2022, https://hbr.org/2022/12/chatgpt-is-a-tipping-point-for-ai.

23. Mollick, "ChatGPT Is a Tipping Point for AI."

24. Kathy Hirsh-Pasek and Elias Blinkoff, "ChatGPT: Educational Friend or Foe?," *Brookings*, June 9, 2023, https://www.brookings.edu/blog/education-plus-development/2023/01/09/chatgpt-educational-friend-or-foe/; Jason Wingard, "ChatGPT: A Threat to Higher Education?," *Forbes*, January 10, 2023, https://www.forbes.com/sites/jasonwingard/2023/01/10/chatgpt-a-threat-to-higher-education/; Phanish Puranam, "ChatGPT and the Future of Business Education," *INSEAD Knowledge*, February 1, 2023, https://knowledge.insead.edu/leadership-organisations/chatgpt-and-future-business-education.

25. Unnati Narang, "The Impact of Generative Artificial Intelligence in Online Learning Platforms," Working Paper, 2023.

26. Katharina Buchholz, "This Is How Much the Global Literacy Rate Grew over 200 Years," *World Economic Forum*, September 12, 2022, https://www.weforum.org/agenda/2022/09/reading-writing-global-literacy-rate-changed/.

27. Buchholz, "This Is How Much the Global Literacy Rate Grew over 200 Years."

28. Buchholz, "This Is How Much the Global Literacy Rate Grew over 200 Years."

29. Buchholz, "This Is How Much the Global Literacy Rate Grew over 200 Years."

30. National Student Clearinghouse Research Center, "Persistence and Retention Fall 2020 Beginning Postsecondary Student Cohort," June 2022, https://nscresearchcenter.org/wp-content/uploads/PersistenceRetention2022.pdf.

31. National Student Clearinghouse Research Center, "Persistence and Retention Fall 2020 Beginning Postsecondary Student Cohort."

32. Lee Gardner, "How A.I. Is Infiltrating Every Corner of the Campus," *Chronicle of Higher Education*, April 8, 2018, https://www.chronicle.com/article/how-a-i-is-infiltrating-every-corner-of-the-campus/.

33. Gardner, "How A.I. Is Infiltrating Every Corner of the Campus."

34. Gardner, "How A.I. Is Infiltrating Every Corner of the Campus."

35. Georgia State University News Hub, "Classroom Chatbot Improves Student Performance, Study Says," press release, March 21, 2022, https://news.gsu.edu/2022/03/21/classroom-chatbot-improves-student-performance-study-says/.

36. Georgia State University News Hub, "Classroom Chatbot Improves Student Performance, Study Says."

37. Ben Gose, "When the Teaching Assistant Is a Robot," *Chronicle of Higher Education*, October 23, 2016, https://www.chronicle.com/article/when-the-teaching-assistant-is-a-robot/.

38. Preston Fore, "Sal Khan Helped Usher in an Era of Online Learning through Khan Academy. Will Its AI tool, Khanmigo, Be a Model for the Future of Education?," *Fortune*, October 4, 2023, https://fortune.com/education/articles/khan-academy-sal-khan-khanmigo-ai-future-of-education/.

39. Carnegie Mellon University Human-Computer Interaction Institute, "HCII Researchers Awarded $2M Grant to Test AI-Based Mobile Tutoring," *Carnegie Mellon University News & Events*, October 18, 2022, https://hcii.cmu.edu/news/hcii-researchers-awarded-2m-grant-test-ai-based-mobile-tutoring-software.

40. Holon IQ, "The 2021 Global Learning Landscape," July 17, 2020, https://www.holoniq.com/notes/the-2021-global-learning-landscape-an-open-source-taxonomy-to-map-the-future-of-education.

41. Jenny C. Aker, Christopher Ksoll, and Travis J. Lybbert, "Can Mobile Phones Improve Learning? Evidence from a Field Experiment in Niger," *American Economic Journal: Applied Economics* 4, no. 4 (October 1, 2012): 94–120, https://doi.org/10.1257/app.4.4.94.

42. Ben York and Susanna Loeb, "One Step at a Time: The Effects of an Early Literacy Text Messaging Program for Parents of Preschoolers," *National Bureau of Economic Research* Working Paper No. w20659, November 1, 2014, https://doi.org/10.3386/w20659.

43. Noam Angrist, Peter Bergman, Caton Brewster, and Moitshepi Matsheng, "Stemming Learning Loss during the Pandemic: A Rapid Randomized Trial of a Low-Tech Intervention in Botswana," *Centre for the Study of African Economies* Working Paper WPS/2020-13, August 2020, https://www.povertyactionlab.org/sites/default/files/research-paper/working-paper_8778_Stemming-Learning-Loss-Pandemic_Botswana_Aug2020.pdf.

44. Zhe Deng, Aaron M. Cheng, Pedro Lopes Ferreira, and Paul A. Pavlou, "From Smart Phones to Smart Students: Distraction versus Learning with Mobile Devices in the Classroom," Social Science Research Network working paper, September 12, 2023, https://doi.org/10.2139/ssrn.4028845.

45. Shelly Culbertson, James Dimarogonas, Katherine Costello, and Serafina Lanna, "Crossing the Digital Divide: Applying Technology to the Global Refugee Crisis," *RAND Corporation*, December 17, 2019, https://www.rand.org/pubs/research_reports/RR4322.html.

46. John F. Pane, Beth Ann Griffin, Daniel F. McCaffrey, and Rita T. Karam, "Effectiveness of Cognitive Tutor Algebra I at Scale," *Educational Evaluation and Policy Analysis* 36, no. 2 (June 1, 2014): 127–144, https://doi.org/10.3102/0162373713507480.

47. Pane et al., "Effectiveness of Cognitive Tutor Algebra I at Scale."

48. Sander Tars, "AI-Driven Personalized Learning Paths: The National Project of Estonia," *MindTitan* (blog), October 21, 2021, https://mindtitan.com/resources/blog/ai-driven-personalized-learning/.

49. Eva Toome, "Estonia to Unleash AI for Personalisation of Education," *Education Estonia*, November 27, 2020, https://www.educationestonia.org/estonia-to-unleash-ai-for-personalisation-of-education/.

50. Toome, "Estonia to Unleash AI for Personalisation of Education."

51. Rachel Sylvester, "How Estonia Does It: Lessons from Europe's Best School System," *Times* Education Commission, January 26, 2022, https://www.thetimes.co.uk/article/times-education-commission-how-estonia-does-it-lessons-from-europe-s-best-school-system-qm7xt7n9s; Organisation for Economic Co-operation and Development, "FAQ," *OECD Programme for International Student Assessment*, n.d., https://www.oecd.org/pisa/pisafaq/.

52. Anuj Kumar and Amit Mehra, "Improving Educational Delivery in K-12 Schools with Personalization Evidence from Randomized Field Experiment in India," Social Science Research Network working paper, April 10, 2024, https://doi.org/10.2139/ssrn.2756059.

53. Kumar and Mehra, "Improving Educational Delivery in K-12 Schools with Personalization."

54. University of Arizona, "Predicting and Enhancing Student Retention with Big Data," News, Eller College of Management, January 30, 2020, https://eller.arizona.edu/news/2017/02/predicting-enhancing-student-retention-big-data.

55. Marisa Garanhel, "5 Real-Life Use Cases of Artificial Intelligence in Education," AI Accelerator Institute, June 17, 2022, https://www.aiacceleratorinstitute.com/5-real-life-use-cases-of-artificial-intelligence-in-education/.

56. Lee Gardner, "How A.I. Is Infiltrating Every Corner of the Campus," *Chronicle of Higher Education*, April 8, 2018, https://www.chronicle.com/article/how-a-i-is-infiltrating-every-corner-of-the-campus/.

57. Matthew Lynch, "26 Ways That Artificial Intelligence (AI) Is Transforming Education for the Better," *The Edvocate*, April 29, 2019, https://www.theedadvocate.org/26-ways-that-artificial-intelligence-ai-is-transforming-education-for-the-better/.

58. Timothy Burke, "Academia: Cheating, Writing and Learning (AI Edition)," *Eight by Seven*, September 30, 2021, https://timothyburke.substack.com/p/academia-cheating-writing-and-learning.

59. Burke, "Academia."

60. David Brooks, "Opinion | In the Age of A.I., Major in Being Human," *New York Times*, February 4, 2023, https://www.nytimes.com/2023/02/02/opinion/ai-human-education.html.

CHAPTER 6

1. Ketki V. Deshpande, Shimei Pan, and James R. Foulds, "Mitigating Demographic Bias in AI-Based Resume Filtering," in *Adjunct Publication of the 28th ACM Conference on User Modeling, Adaptation and Personalization* (UMAP '20 Adjunct) (New York: Association for Computing Machinery, 2020), 268–275, https://doi.org/10.1145/3386392.3399569.

2. Liz Ryan, "No, You're Not Crazy—The Hiring Process Is Broken," *Forbes*, April 12, 2018, https://www.forbes.com/sites/lizryan/2018/04/12/no-youre-not-crazy-the-hiring-process-is-broken/.

3. Karsten Strauss, "The Role of Artificial Intelligence in the Future of Job Search," *Forbes*, February 2, 2018, https://www.forbes.com/sites/karstenstrauss/2018/02/02/the-role-of-artificial-intelligence-in-the-future-of-job-search/.

4. SHRM, "Interactive Chart: How Historic Has the Great Resignation Been?," Society for Human Resource Management, April 20, 2022, https://www.shrm.org/resourcesandtools/hr-topics/talent-acquisition/pages/interactive-quits-level-by-year.aspx.

5. Erik Brynjolfsson, Danielle Li, and Lindsey Raymond, "Generative AI at Work," National Bureau of Economic Research Working Paper No. w31161, April 1, 2023, https://doi.org/10.3386/w31161.

6. Gergo Vari, "How to Use AI-Based Tools to Find Your Next Job," *Fast Company*, February 7, 2022, https://www.fastcompany.com/90719364/how-to-use-ai-based-tools-to-find-your-next-job.

7. Strauss, "The Role of Artificial Intelligence in the Future of Job Search."

8. Jose Maria Barrero, Nicholas Bloom, and Steven J. Davis, "Why Working from Home Will Stick," National Bureau of Economic Research Working Paper No. w28731, April 1, 2021, https://doi.org/10.3386/w28731.

9. Alana Rudder, "The Future of Work Is Now: AI Helps Job Hunters Find, Land, & Keep Dream Jobs," *Medium*, July 5, 2018, https://towardsdatascience.com/the-future-of-work-is-now-ai-helps-job-hunters-find-land-keep-dream-jobs-b58a3a247c34.

10. Bernard Marr, "The Amazing Ways How Unilever Uses Artificial Intelligence to Recruit & Train Thousands of Employees," *Bernard Marr* (blog), July 13, 2021, https://bernardmarr.com/the-amazing-ways-how-unilever-uses-artificial-intelligence-to-recruit-train-thousands-of-employees/.

11. Bernard Marr, "Job Search in the Age of Artificial Intelligence—5 Practical Tips," *Bernard Marr* (blog), July 13, 2021, https://bernardmarr.com/job-search-in-the-age-of-artificial-intelligence-5-practical-tips/.

12. Maxime Legardez Coquin, "Hiring Is Broken—Here's How We Can Fix It," *Forbes*, June 22, 2022, https://www.forbes.com/sites/forbeshumanresourcescouncil/2022/06/22/hiring-is-broken-heres-how-we-can-fix-it/.

13. Ishita Chakraborty, Khai Chiong, Howard Dover, and K. Sudhir, "AI and AI-Human Based Salesforce Hiring Using Interview Videos," Social Science Research Network working paper, April 7, 2023, https://doi.org/10.2139/ssrn.4137872.

14. Chakraborty et al., "AI and AI-Human Based Salesforce Hiring."

15. Chakraborty et al., "AI and AI-Human Based Salesforce Hiring."

16. Joseph Fuller, "Skills-Based Hiring Is on the Rise," *Harvard Business Review*, December 2, 2022, https://hbr.org/2022/02/skills-based-hiring-is-on-the-rise.

17. Nik Dawson, Mary-Anne Williams, and Marian-Andrei Rizoiu, "Skill-Driven Recommendations for Job Transition Pathways," *PLOS ONE* 16, no. 8 (August 4, 2021): e0254722, https://doi.org/10.1371/journal.pone.0254722.

18. Sonia K. Kang, Katherine A. DeCelles, András Tilcsik, and Sora Jun, "Whitened Résumés," *Administrative Science Quarterly* 61, no. 3 (July 8, 2016): 469–502, https://doi.org/10.1177/0001839216639577.

19. Marianne Bertrand and Sendhil Mullainathan, "Are Emily and Greg More Employable Than Lakisha and Jamal? A Field Experiment on Labor Market Discrimination," *American Economic Review* 94, no. 4 (August 1, 2004): 991–1013, https://doi.org/10.1257/0002828042002561.

20. Corinne A. Moss-Racusin, John F. Dovidio, Victoria L. Brescoll, Mark Graham, and Jo Handelsman, "Science Faculty's Subtle Gender Biases Favor Male Students," *Proceedings of the National Academy of Sciences* 109, no. 41 (September 17, 2012): 16474–16479, https://doi.org/10.1073/pnas.1211286109. See also Anja Lambrecht and Catherine Tucker, "Algorithmic Bias? An Empirical Study of Apparent Gender-Based Discrimination in the Display of STEM Career Ads," *Management Science* 65, no. 7 (July 1, 2019): 2966–2981, https://doi.org/10.1287/mnsc.2018.3093.

21. David Neumark, Ian Burn, and Patrick Button, "Age Discrimination and Hiring of Older Workers," *Federal Reserve Bank of San Francisco* 2017–06 (February 27, 2017),

https://www.frbsf.org/economic-research/publications/economic-letter/2017/february/age-discrimination-and-hiring-older-workers/.

22. Devah Pager and Lincoln Quillian, "Walking the Talk? What Employers Say versus What They Do," *American Sociological Review* 70, no. 3 (June 2005): 355–380, https://www.jstor.org/stable/4145386.

23. Frida Polli, "Using AI to Eliminate Bias from Hiring," *Harvard Business Review*, January 18, 2023, https://hbr.org/2019/10/using-ai-to-eliminate-bias-from-hiring.

24. Danielle Li, Lindsey R. Raymond, and Peter Bergman, "Hiring as Exploration," National Bureau of Economic Research Working Paper No. w27736, August 1, 2020, https://doi.org/10.3386/w27736.

25. Edward W. Felten, Manav Raj, and Robert Seamans, "A Method to Link Advances in Artificial Intelligence to Occupational Abilities," *AEA Papers and Proceedings* 108 (2018): 54–57.

26. Shakked Noy and Whitney Zhang, "Experimental Evidence on the Productivity Effects of Generative Artificial Intelligence," Social Science Research Network working paper, March 1, 2023, https://ssrn.com/abstract=4375283.

27. Erdem Dogukan Yilmaz, Ivana Naumovska, and Vikas A. Aggarwal, "AI-Driven Labor Substitution: Evidence from Google Translate and ChatGPT," INSEAD Working Paper No. 2023/24/EFE, March 26, 2023, https://ssrn.com/abstract=4400516.

28. Hongxian Huang, Runshan Fu, and Anindya Ghose, "The Economic Impact of Adopting Generative AI: A Multi-Platform Analysis," Social Science Research Network working paper, December 20, 2023, https://papers.ssrn.com/sol3/papers.cfm?abstract_id=4670714.

29. Kartik Hosanagar, "Using ChatGPT for Market Research?," LinkedIn blog post, September 15, 2023, https://www.linkedin.com/pulse/using-chatgpt-market-research-kartik-hosanagar/.

30. John Scott Lewinski, "HAL's Pals: Top 10 Evil Computers," *WIRED*, January 9, 2009, https://www.wired.com/2009/01/top-10-evil-com/.

31. AI Now Institute, "About Us," June 7, 2023, https://ainowinstitute.org/about. AI Now Institute focuses on examining AI's social implications.

32. Drew Harwell, "A Face-Scanning Algorithm Increasingly Decides Whether You Deserve the Job," *Washington Post*, November 6, 2019, https://www.washingtonpost.com/technology/2019/10/22/ai-hiring-face-scanning-algorithm-increasingly-decides-whether-you-deserve-job/; Drew Harwell, "Rights Group Files Federal Complaint against AI-Hiring Firm HireVue, Citing 'Unfair and Deceptive' Practices," *Washington Post*, November 6, 2019, https://www.washingtonpost.com/technology/2019/11/06/prominent-rights-group-files-federal-complaint-against-ai-hiring-firm-hirevue-citing-unfair-deceptive-practices/.

33. Chamorro-Premuzic, Tomas, and Reece Akhtar, "Should Companies Use AI to Assess Job Candidates?," *Harvard Business Review*, May 17, 2019, https://hbr.org/2019/05/should-companies-use-ai-to-assess-job-candidates.

34. Julia Angwin, Jeff Larson, Surya Mattu, and Lauren Kirchner, "Machine Bias," *ProPublica*, May 23, 2016, https://www.propublica.org/article/machine-bias-risk-assessments-in-criminal-sentencing.

35. Dastin, "Amazon Scraps Secret AI Recruiting Tool."

36. Polli, "Using AI to Eliminate Bias from Hiring."

37. Chamorro-Premuzic and Akhtar, "Should Companies Use AI to Assess Job Candidates?"

38. Harwell, Face-Scanning Algorithm."

39. Polli, "Using AI to Eliminate Bias from Hiring"; Matt Gonzales, "AI-Based Bias a Hot Topic of Discussion during EEOC-Led Meeting," *SHRM*, October 7, 2022, https://www.shrm.org/resourcesandtools/hr-topics/behavioral-competencies/global-and-cultural-effectiveness/pages/ai-based-bias-a-hot-topic-of-discussion-during-eeoc-led-meeting.aspx.

40. Bo Cowgill, "Bias and Productivity in Humans and Algorithms: Theory and Evidence from Résumé Screening," 2018, https://www.semanticscholar.org/paper/Bias-and-Productivity-in-Humans-and-Algorithms%3A-and-Cowgill/11a065f86b549892a01388bb579cdc2bf4165dca.

41. Jennifer Kirkwood, "New AI HR/Talent Laws Get the Attention of the C-Suite," *IBM Blog*, April 4, 2023, https://www.ibm.com/blog/new-ai-hr-talent-laws-get-the-attention-of-the-c-suite/.

42. Noah Smith, "A Job Is More Than a Paycheck," *Bloomberg.com*, December 2, 2016, https://www.bloomberg.com/opinion/articles/2016-12-02/a-job-is-more-than-a-paycheck.

43. Jeffrey Pfeffer, *Dying for a Paycheck: How Modern Management Harms Employee Health and Company Performance—and What We Can Do About It* (New York: HarperCollins, 2018), https://www.harpercollins.com/products/dying-for-a-paycheck-jeffrey-pfeffer.

44. Tom Rath, "I've Spent 2 Decades Studying How Work Affects Our Health and Well-Being, and One Solution Is Clear: Your Job Has to Serve a Purpose beyond a Paycheck," *Business Insider*, February 4, 2020, https://www.businessinsider.com/work-should-have-purpose-beyond-paycheck-health-well-being.

CHAPTER 7

1. Saturday Night Live, "Amazon Echo—SNL," May 14, 2017, https://www.youtube.com/watch?v=YvT_gqs5ETk.

2. Wikipedia contributors, "The Jetsons," *Wikipedia*, May 17, 2023, https://en.wikipedia.org/wiki/The_Jetsons.

3. Google, "Nest Learning Thermostat," n.d., https://store.google.com/product/nest_learning_thermostat_3rd_gen.

4. Eight Sleep, "Pod Intelligent Cooling & Heating Mattress," n.d., https://www.eightsleep.com/pod-mattress/.

5. Google, "Nest Learning Thermostat."

6. Larry Dignan, "How Best Buy Plans to Expand into Home Healthcare Services, Remote Monitoring to Help Seniors Age in Place," *ZDNET*, September 30, 2019, https://www.zdnet.com/article/how-best-buy-plans-to-expand-into-home-healthcare-services-remote-monitoring-to-help-seniors-age-in-place/.

7. Anindya Ghose, *Tap: Unlocking the Mobile Economy* (Cambridge, MA: MIT Press), 39.

8. Sarah Banks, "A Historical Analysis of Attitudes Toward the Use of Calculators in Junior High and High School Math Classrooms in the United States Since 1975," Master of Education Research Thesis, Cedarville University School of Education, 2011, https://digitalcommons.cedarville.edu/education_theses/31/.

9. Jennifer Breheny Wallace, "Instagram Is Even Worse than We Thought for Kids. What Do We Do about It?," *Washington Post*, September 21, 2021. https://www.washingtonpost.com/lifestyle/2021/09/17/instagram-teens-parent-advice/.

10. Amelia Hill, "Voice Assistants Could 'Hinder Children's Social and Cognitive Development,'" *Guardian*, September 28, 2022, https://www.theguardian.com/technology/2022/sep/28/voice-assistants-could-hinder-childrens-social-and-cognitive-development.

11. Samantha Murphy Kelly, "Growing Up with Alexa: A Child's Relationship with Amazon's Voice Assistant," *CNN*, October 16, 2018, https://www.cnn.com/2018/10/16/tech/alexa-child-development/index.html.

12. Sarah McQuate, "Do Alexa and Siri Make Kids Bossier? New Research Suggests You Might Not Need to Worry," *UW News*, September 13, 2021, https://www.washington.edu/news/2021/09/13/alexa-siri-make-kids-bossier-research-suggests-you-might-not-need-to-worry/.

13. Erin Beneteau, Ashley Boone, Yuxing Wu, Julie A. Kientz, Jason C. Yip, and Alexis Hiniker, "Parenting with Alexa: Exploring the Introduction of Smart Speakers on Family Dynamics," *CHI '20: Proceedings of the 2020 CHI Conference on Human Factors in Computing Systems* (April 2020): 1–13, https://doi.org/10.1145/3313831.3376344.

14. Alexis Hiniker, Amelia Wang, Jonathan A. Tran, Mingrui Zhang, Jenny Radesky, Kiley Sobel, and Sung-Soo Hong, "Can Conversational Agents Change the Way Children Talk to People?," *IDC '21: Interaction Design And Children* (June 24, 2021): 338–349, https://doi.org/10.1145/3459990.3460695.

15. McQuate, "Do Alexa and Siri Make Kids Bossier?"

CHAPTER 8

1. Bill Gates [@billgates], Twitter post, March 21, 2023, 11:09 a.m., https://twitter.com/BillGates/status/1638226408412708864.

2. Ron Kohavi and Stefan Thomke, "The Surprising Power of Online Experiments., *Harvard Business Review*, September 16, 2017, https://hbr.org/2017/09/the-surprising-power-of-online-experiments.

3. Varun Gulshan, Lily Peng, Marc Coram, Martin C. Stumpe, Derek Wu, Arunachalam Narayanaswamy, Subhashini Venugopalan, et al., "Development and Validation of a Deep Learning Algorithm for Detection of Diabetic Retinopathy in Retinal Fundus Photographs," *JAMA* 316, no. 22 (December 13, 2016): 2402, https://doi.org/10.1001/jama.2016.17216.

4. Patrick M. Heck, Christopher F. Chabris, Duncan J. Watts, and Michelle L. Meyer, "Objecting to Experiments Even While Approving of the Policies or Treatments They Compare," *Proceedings of the National Academy of Sciences of the United States of America* 117, no. 32 (July 27, 2020): 18948–18950, https://doi.org/10.1073/pnas.2009030117.

5. "ImageNet," Stanford Vision Lab, Stanford University, Princeton University, n.d., https://image-net.org/; Dave Gershgorn, "The Data That Transformed AI Research—and Possibly the World," *Quartz*, July 20, 2022, https://qz.com/1034972/the-data-that-changed-the-direction-of-ai-research-and-possibly-the-world/.

6. Alex Krizhevsky, Ilya Sutskever, and Geoffrey E. Hinton, "ImageNet Classification with Deep Convolutional Neural Networks," in *Advances in Neural Information Processing Systems 25 (NIPS 2012)*, ed. F. Pereira, C. J. Burges, L. Bottou, and K. Q. Weinberger, NeurIPS Proceedings, 2012, https://papers.nips.cc/paper_files/paper/2012/hash/c399862d3b9d6b76c8436e924a68c45b-Abstract.html.

7. Kate Crawford and Trevor Paglen, "Excavating AI: The Politics of Images in Machine Learning Training Sets," n.d., https://www.excavating.ai/.

8. Julia Angwin, Jeff Larson, Surya Mattu, and Lauren Kirchner, "Machine Bias," *ProPublica*, May 23, 2016, https://www.propublica.org/article/machine-bias-risk-assessments-in-criminal-sentencing.

9. Craig S. Smith, "Dealing with Bias in Artificial Intelligence," *New York Times*, January 3, 2020, https://www.nytimes.com/2019/11/19/technology/artificial-intelligence-bias.html.

CONCLUSION

1. Mayo Clinic Equity, Inclusion and Diversity, August 11, 2023, Fireside chat discussing #AI, #Platform, #MachineLearning with @jhalamka @ravibapna, and Sonya Makhni, M.D., M.S. Exploring the impact of AI, using AI algorithm insights, and monitoring for bias, at https://x.com/MayoEquity/status/1690008344063279104?s=20.

2. See https://business.purdue.edu/events/data4good/home.php for details.

3. Marc Andreessen [@pmarca], "Why AI Will Save the World," Twitter, February 8, 2022, 11:11 p.m., https://twitter.com/pmarca/status/1666112323713662977?s=46.

4. Aryamala Prasad, "Unintended Consequences of GDPR: A Two-Year Lookback," Regulatory Studies Center, Trachtenberg School of Public Policy & Public Administration, Columbian

College of Arts & Sciences, George Washington University, September 3, 2020, https://regulatorystudies.columbian.gwu.edu/unintended-consequences-gdpr.

5. Prasad, "Unintended Consequences of GDPR."

6. David Brooks, "Opinion: In the Age of A.I., Major in Being Human," *New York Times*, February 4, 2023, https://www.nytimes.com/2023/02/02/opinion/ai-human-education.html.

ACKNOWLEDGMENTS

1. S. K. Kulkarni, "Obituary: Late Professor Jawahar Singh Bapna (1942–2021)," *Indian Journal of Pharmacology* 53, no. 5 (2021): 423, https://www.ijp-online.com/text.asp?2021/53/5/423/331091.

Index

A/B testing, 59
AdmitHub, 99
African Americans
  discrimination toward, 57
  medical patients, 75
AI optimists and optimism, 3, 156–157
Airbnb, 15, 46–51, 60
  discrimination and, 57–59
  negative impacts of, 51–53
  and trust, 53–56
"Airbnb Pits Neighbor Against Neighbor in Tourist-Friendly New Orleans" (*New York Times*), 51–52
AI summit (authors' metaphor), 141–143
  and the four pillars in the House of AI, 18–21, 69, 145–148
  leadership and organizational complements, 149–150
  and patterns of confusion, 144–145
Aker, Jenny C., 101
Akhtar, Reece, 122
Aleven, Vincent, 100–101
Alex (Jasmina's fictional friend, the triathlete), 8, 9, 66–67, 72
Alexa (voice-activated assistant), 14, 127
Algorithmic bias, 123
Algorithms of Oppression: How Search Engines Reinforce Racism (Noble), 3
AlphaFold, 82
ALS, treatment of, 83
Altman, Sam, 156
Alyssa (Jasmina's fictional friend), 10–12, 14, 15, 16–18, 24–25, 28–29
Amazon, 9, 16, 59
  AI-based recruiting process, 109
Amazon Marketplace, 56
American Heart Association, 75
Andreessen, Marc, 157
Android, emergency location service (ELS), 4
Anomaly detection, 8–10
Apple Watch, 5, 14, 64, 67, 68, 71
Artificial general intelligence (AGI), 6
Artificial intelligence (AI). *See also* ChatGPT; Large language models (LLMs); Machine learning
  and an AI-ready workforce, 150–151
  fears about, 121–123
  idea of, 87
  optimism, 3, 156–157
  strong, 6
  weak, 6
Atlantic, The (magazine), 90
Attention mapping, 93–94
Augmedix, 78
Australia, 117–118
Australian Bureau of Statistics (ABS), 117
Authentication, and mobile dating apps, 35–38

Banerji, Abhijit, 16
Bannister, Roger, 149
Bapna, Ravi, 14, 21, 32–33, 45–48, 50, 68, 76, 81, 88–89, 96, 141, 150
Battlestar Galactica (film), 87
Beneteau, Erin, 137
Best Buy, 17–18, 134
Bias and biases
  algorithmic, 151–152
  in recruiting and hiring, 118–121, 125
Bidirectional recurrent neural networks (BRNNs), 45
Black Mirror (TV series), 87
Blade Runner (film), 3
Blinkoff, Elias, 89, 91, 95
BodyMedia, 69
Bogost, Ian, 90
Boston University, 15
Botswana, 101
Brockman, Greg, 87–88
Brooks, David, 158–159
Buddha, 159
Bumble, 9, 35, 38–40
Burke, Timothy, 105
Burtch, Gordon, 88–89, 96
Butler, Samuel, 87

Cambridge Analytica, 3–4
Cancer, in white versus Black populations, 5
Candy Crush, 37
Career fulfillment, and AI, 123–125
Carlos (who has experience as a restaurant inspector), 117, 118
Carnegie Mellon University, 15, 100–101
Casio sports watch, 68
Causal analytics, 19–20, 21, 78, 146–147
Causal inference, 7–8
  anomaly detection, 8–10
Chamorro-Premuzic, Tomas, 122
ChatGPT (Chat Generative Pre-trained Transformer), 87–91, 156, 157
  aspects of method, 91–96
  Copilot, 2
  ChatGPT-3, 93
  ChatGPT-4, 6–7, 15, 45, 93–94, 110, 158
Chesky, Brian, 50–51
Children, and AI technology, 136–138
Chronicle of Higher Education, The (trade journal), 98, 99
Clarke, Arthur C., 121
Cognitive behavioral therapy (CBT), 75–76, 85
Comfort, and AI, 130
Common Crawl, 92
COMPAS, 152
Consumer Reports (magazine), 54
Conversational agent, 137
Copilot, 2
Coquin, Maxime Legardez, 115, 116
Correlation, 146–147
Correlational mapping of inputs to outputs, 146–147
Cortana, 127
Counselor, The, 3
COVID-19, 4, 40, 60
  vaccine, 83
Cowgill, Bo, 123
Coy, Peter, 95
Crawford, Kate, 152
Customer Relationship Management (CRM), 144
CYBER (podcast), 3

Daily Show, The (TV show), 78
Damiani, Marcello, 83–84
Data engineering, 145
Data Science University (Optum), 81
Dating, online AI, 24–25
  and anonymity, 32–35
  effect of, 28–31
  good, bad, or indifferent, 31–32
  and metaverse, 40
  and natural language processing, 25–28
DeepETA, 53
Deep learning, 6, 14–15, 152

Deng, Zhe, 101–102
Descriptive and predictive analytics, 19, 20, 145–146
Discrimination
  and Airbnb, 57–59
  in hiring, 118–121
District of Columbia v. Facebook Inc., 4
Doctor Who (TV series), 121
Dying for a Paycheck: How Modern Management Harms Employee Health and Company Performance—and What We Can Do About It (Pfeffer), 124

eBay, 60
ECGs (echocardiograms), 6
Education, AI and ML in, 96, 105–106
  answering students' routine questions, 98–100
  and literacy, 97
  and personalization 102–103
  and post-secondary education, 97–98
  predicting student retention, 103–105
  tutors, 101–102
Education Estonia (trade journal), 103
eharmony, 30–31
Electronic health record (EHR), 72, 80
Electronic recordkeeping, 72
Eller College (University of Arizona), 104
Embedding, 27, 92
Emergency location service (ELS), 4
Enterprise Resource Planning (ERP), 144
Entertainment, and AI, 129–130
EPIC, 155–156
Erewhon (Butler), 87
Erwin (fictional Tinder user), 37
Esteril Conferences (2023), 156
Estonia, 102–103
Etsy, 60
Excavating AI: The Politics of Images in Machine Learning Training Sets (Crawford and Paglen), 152
Excel, and ChatGPT, 2

Facebook, 4, 16, 38, 55, 56
Fast Company (Vari), 113
Fauci, Anthony, 156
Fears, about AI, 121–123
Fellbaum, Christiane, 151
Felten, Edward W., 119–120
Financial Times (newspaper), 81, 82–83
FIT Armbands, 64, 69
Fiverr, 58
Flywheel, 57
Forbes (magazine), 110, 113
Fourth Industrial Revolution, 97
Frankenstein (Shelley), 87
Friar, Sarah, 43–44
Fu, Runshan, 120

Gamification, of mobile dating apps, 35–38
Gebbia, Joe, 50–51
Geek Squad, Best Buy, 134
General Data Protection Regulation (GDPR), 158
Generative AI, 1–2, 100, 119–120, 125, 148
Georgia Institute of Technology, 99–100
Georgia State University, 98–99
Getaround, 46
Ghose, Anindya, 4, 66, 68–69, 72, 76–77, 120, 135, 136, 141, 150, 156–157
Gmail, 25
Goel, Ashok K., 99–100
Good Housekeeping Seal of Approval, 54
Goodreads, 9
Google, 9, 59
  Nest Thermostat, 132–133
Google Assistant, 127
Google Glass, 78
  Glass Enterprise Edition, 78
GQ (magazine), 39
Graduate Record Examinations (GRE), and ChatGPT, 93–94
Graffiti Kingdom, 120
Great Resignation of 2021, 110–111, 115
Grindr, 35

Habeler, Peter, 149
Halamka, John, 155
Hanna-Barbera, 128
Harvard Business Review, 90, 95, 116, 119
Harvard University, 15
Hazel (fictional parent of newborn), 63–66, 80
Health literacy, 73–74
Heart rate variability (HRV), 64–66
Hersh, Eitan, 4
Hillary, Edmund, 142–143
Hinton, Geoffrey, 3, 70, 87, 152
HireVue, 121–123
Hiring, 112–113
  benefits of AI for job seekers, 113–116
  biases in, 118–121, 125
  career fulfillment, 123–125
  evolving work landscape, 109–112
  job satisfaction 123–125
  salesforce, 115–116
  skills-based, 116–118
Hirsh-Pasek, Kathy, 89, 91
Homes. *See* Smart homes
Hosanagar, Kartik, 121
House of AI framework, 18–21, 69, 145–148
Huang, Hongxian, 120
Human-Computer Action Institute (Carnegie Mellon University), 100–101
Hunt, John, 142
Hurff, Scott, 37–38
Hyperspace, 10

IBM, 123
Image classification algorithms, 152
ImageNet, 151–153
  Roulette project, 153
Indeed.com, 110
India, 45–46, 103
Informational loss, 72
Instagram, 16
Intent analytics, 48–50
"In the Age of A.I., Major in Being Human" (Brooks), 158–159

Jasmina (fictional single mother), 7–11, 12, 15, 18, 19, 23, 24–25, 26, 27, 28–29, 116, 118
  and hiring bias, 107–108, 109, 111
Jawahar (Ravi's father), 47–48, 50, 53
JawBone, 69
Jetsons, The (cartoon), 128
  George Jetson (cartoon character), 128
Jill Watson (bot), 99–100
Job satisfaction, and AI, 123–125
Jordan, DeAndre, 64

Khan Academy, 100
Khanmigo (personalized tutor), 100
Kompella, Kashyap, 76
Kubrick, Stanley, 121

Large language models (LLMs), 6–7, 91–92, 94, 95. *See also* ChatGPT
  and bias in hiring, 120
  LLM-based teaching assistants, 100
Larsen, Loren, 122–123
Latent Dirichlet allocation (LDA), 49
Leaders and leadership, 143
LeCun, Yann, 3
LDL cholesterol, 14
LGBTQ+ individuals
  and Airbnb, 57
  and bias in hiring, 118–119
Li, Fei-Fei, 151
Lighting, and AI, 130–131
LinkedIn, 16, 55, 121
Loeb, Susanna, 101
Long short-term memory (LSTM), 45
Love, finding, 23–24. *See also* Dating, online AI
Lyft, 53, 54, 57, 58–59, 60
Lytvyn, Max, 25

Machine learning (ML), 1, 7–8, 44, 146. *See also* Causal inference
  deep learning, 6, 14–15, 152
  reinforcement learning, 16–17

supervised machine learning algorithms, 11–14
unsupervised machine learning algorithms, 8–10
Malgieri, Gianclaudio, 1
MaMMa Klinika (Hungary), 70
Marcus, Gary, 90
Market research, and LLM tools, 120–121
Marr, Bernard, 114–115
Mashable (news site), 78
Massive open online courses (MOOCs), 100
Match.com, 24
Matrix, The (film), 3, 87
Mattis, James, 156
Max (who knows how to fix automobiles), 117, 118
Mayo Clinic, RISE Conference (2023), 155
Mayo Clinic Platform, 155
Meja, Jorge, 57
Mental Health American, 75
Messner, Reinhold, 149
Meta, 59
Mhealth (mobile health), 63–66, 67–71, 84–85
mental health and digitized therapy, 75–76
mobile devices, 71–73
and personalization, 76–78
prediction and prevention, 80–82
Microsoft, 59, 149
Moderna, 83–84
Mollick, Ethan, 90, 95
Mount Everest, 141–143
“mRNA-4157—Personalized Cancer Vaccine,” 83–84
Mukherjee, Raj, 110

Nadella, Satya, 2
Narang, Unnati, 96
National Bureau of Economics Research, 111–112
National Health Care Anti-Fraud Association (NHCAA), 81
Natural language processing (NLP), 25–28, 44–45
Nature (journal), 3, 82
Negativity bias, 53
Nest Thermostat, 132, 139
Netflix, 16, 129
New Yorker, The (magazine), 70
New York Times, The (newspaper), 1, 51, 70, 90, 95, 158
Nextdoor, 43–44
Niger, 101
Noy, Shakked, 120

OKCupid (dating app), 6, 9, 26, 35, 50
OpenAI, 78, 94. *See also* ChatGPT
Organization for Economic Co-operation and Development (OECD), 103
Outcomes, machine learning, 12
Overgoor, Jan, 59

Paglen, Trevor, 152
Parker, Chris, 57
Pasquale, Frank, 1
Pattern mining, 146–147
Personalization, 102–103
and mobile health, 76–78
tutor, 100
Pfeffer, Jeffrey, 124
Philips Hue, 130–131
Photos, and mobile dating apps, 35–38
Pinterest, 55
Polli, Frida, 119, 122
Positional encoding, and ChatGPT, 92
PowerPoint, and ChatGPT, 2
Prediction
and health, 80–82
and student retention, 103–105
Predictive analytics, 19, 21
Prescriptive analytics, 20–21, 147–148
Prevention, and health, 80–82
Privacy, and AI, 1, 135–138
Privacy Project, The (*New York Times*), 1

Programme for International Student Assessment (PISA), 103
Prompt engineering, 121
ProPublica (newsroom), 122
Purdue University, 156
Pythagoras, 12
Pythagorean theorem, 8

Quantified self, 67

Rad, Sean, 37
Ram, Sudha, 104
Rath, Tom, 124
Reddit, 37
Reinforcement learning algorithms, 16–17, 119–120
Renick, Timothy N., 98
Research, medical, 82–84
Resnick, Paul, 56
Retention, predicting student, 103–105
Return on investment (ROI) calculations, 20
RISE Conference (2023), 155
RLHF (reinforcement learning with human feedback), 94
Romer, Paul, 156
Roomba, 133
Roose, Kevin, 90, 91
Ross, Maurice, A., 4
Roulette project (ImageNet), 153
Russakovsky, Olga, 152–153
Ryan (fictional dating app user), 34
Ryan, Liz, 110

Sampo, 81
Saturday Night Live (TV show), 78
Scientific American (magazine), 90
Security, home, 131–132
Sharing Economy: The End of Employment and the Rise of Crowd-Based Capitalism, The (Sundararajan), 55
Shelley, Mary, 87
Signaling, in online dating, 33, 34
Siri, 14, 127
Skills-based hiring, 116–118
Skinner, B. F., 37
Smarter Workforce Institute (IBM), 123
Smart homes, 128, 138–139
  and children, 136–138
  and convenience, 133–135
  efficiency in, 132–133
  lighting in, 130–131
  and privacy, 135–138
  security, 131–132
Smartphones, 71, 72. *See also* Mhealth
Smart-pricing algorithm, 46
Smart watches and wristbands, 68. *See also* Mhealth
Snapchat, 37, 55
Social friction, 31, 33
Society for Human Resource Management, 111
Sofia (Ravi's wife), 45–48, 50
Spotify, 129, 130
Star Trek (TV series), 121
Stemming, 26
Stern, Jacob, 90
Sundararajan, Arun, 55, 56
Supervised machine learning algorithms, 11–14
Susan (fictional dating app user), 34
Swiping (Tinder), 36–38
Synapse (algorithm), 31
Synthetic dataset, 148–149

TAP: Unlocking the Mobile Economy (Ghose), 136
TaskRabbit, 54, 55, 58
Telemetric clothing, 68
Tenzing Norgay, 142
Terminator, The (film), 87, 121, 135
Tesla, 1
Text-to-numbers trick, 26, 44
Text-to-numbers-to-embedding trick, 28
Therapeutics, research, 82–84

TikTok, 16, 55, 59
Time (magazine), 39, 40, 43
TIME100 Most Influential Companies of 2022, 43
Tinder, 25, 35–38
Tokenizing, and ChatGPT, 92
Transfer learning, 148–149
Transformer-based approach, and ChatGPT, 92–93
Transformers (film), 135
Tripadvisor, 55
Trust, 56
Turing Award, 3
Turing Test, 1–2
Turo, 46
Tutors and tutoring, AI-based, 100–102
Twitter (X), 16, 55, 59
2001: A Space Odyssey (film), 121

Uber, 53, 57, 58–59, 60
Unilever, 114–115
United States
  Bureau of Labor Statistics, 111
  Federal Communications Commission, 5
  Senate Judiciary Committee, 4
University of Arizona, 104–105
University of Maryland, 77
Unsupervised machine learning algorithms, 8
Upper Confidence Bound (algorithm), 119
Usha (Ravi's mother), 47–48, 50, 53

Vacuum cleaners, robot, 133
Vari, Gergo, 113
Verghese, Abraham, 155
Virtual reality (VR), 40
Voice-activated assistants, 127–128
Voice-recognition technology, 127–128
VRBO, 46

Watson (IBM technology), 99
Weak AI, 6. *See also* Generative AI
Weak signaling, in online AI dating, 34
Weapons of Math Destruction: How Big Data Increases Inequality and Threatens Democracy (O'Neil), 3
Wearables, 68, 78–79. *See also* Mhealth
We Met in Virtual Reality (documentary), 40
Wharton School of Business, 120–121
WHOOP fitness band, 64–66, 80
Wikipedia, 45, 92
Windows, and ChatGPT, 2
Wolfe Herd, Whitney, 38–40
Women, biases in hiring, 118–119
Word, and ChatGPT, 2
Work. *See* Hiring
Workforce, AI-ready, 150–151
World Economic Forum, 97
World Health Organization (WHO), 75
World Knowledge Forum (2023), 156
Wozniak, Steve, 156

X (formerly known as Twitter), 16, 55, 59, 92

Yelp, 55
Yilmaz, Erdem Dogukan, 120
York, Ben, 101
YouTube, 16, 54–55

Zeckhauser, Richard, 56
Zhang, Whitney, 120
Zuckerberg, Mark, 40